The Kingdom
of Infinite Number

The Kingdom
of Infinite Number

A Field Guide

Bryan Bunch

W. H. Freeman and Company
New York, New York

Text Designer: Diana Blume

Credits: *p. vi, top*, From *W. H. Auden: Collected Poems* by *W. H. Auden*, edited by Edward Mendelson. © 1944 and renewed 1972 by W. H. Auden. Reprinted by permission of Random House, Inc. *p. vi, bottom*, From Karl Menninger and Paul Broneer (translator), *Number Words and Number Symbols: A Cultural History of Numbers* (Cambridge, MA: The MIT Press, 1969); *p. vii*, From Peter Høeg, *Smilla's Sense of Snow*, translated by Tiina Nunnally (New York: Farrar, Straus and Giroux, 1993).

Library of Congress Cataloging-in Publication Data

    Bunch, Bryan H.
    The kingdom of infinite number: a field guide / Bryan Bunch.
    p. cm.
    Includes bibliographical references and index.
    ISBN 0-7167-3388-9
    1. Numeration. 2. Number concept I. Title

        QA141 .B76 2000
        513–dc21   99–053641

Printed in the United States of America

First printing 1999

W. H. Freeman and Company
41 Madison Avenue, New York, NY 10010
Houndmills, Basingstoke RG21 6XS, England

For my brother, Dale,
who ponders the large ideas

Great is Caesar: He has conquered Seven Kingdoms.
The Third was the Kingdom of Infinite Number:
Last night it was Rule-of-Thumb, to-night it is To-a-T;
Instead of Quite-a-lot, there is Exactly-so-many;
Instead of Only-a-few, there is Just-these.
Instead of saying, 'You must wait until I have counted,'
We say, 'Here you are. You will find this answer correct;'
Instead of a nodding acquaintance with a few integers,
The Transcendentals are our personal friends.
Great is Caesar: God must be with Him.

W. H. AUDEN, *For the Time Being*

If we address ourselves directly to our number words, "one, two, three . . . nine, ten, hundred, thousand," they remain strangely silent about their inner meanings; in this respect they behave like but few other words of our language. It is remarkable that in our daily lives we encounter them constantly and use them as the most reliable bearers of concepts in our possession. Yet we allow ourselves to be content with their usefulness while knowing them only "by name." They pass by us mute, like alien slaves valued only for their services, and we do not dignify them by inquiring into their "person" or their homeland. And yet they do have "personalities," chosen by early man out of his colorful, chaotic environment to be the bearers of his concepts of numbers.

KARL MENNINGER, *Number Words and Number Symbols*

The foundation of mathematics is numbers. If anyone asked me what makes me truly happy, I would say: numbers. . . . Because the number system is like human life. First you have the natural numbers. The ones that are whole and positive. The numbers of a small child. But human consciousness expands. The child discovers a sense of longing, and do you know what the mathematical expression is for longing? . . . The negative numbers. The formalization of the feeling that you are missing something. And the human consciousness expands and grows even more, and the child discovers the in-between spaces. Between stones, between pieces of moss on the stones, between people. And between numbers. And do you know what that leads to? It leads to fractions. Whole numbers plus fractions produce rational numbers. And human consciousness doesn't stop there. It wants to go beyond reason. It adds an operation as absurd as the extraction of roots. And produces the irrational numbers. . . . It's a form of madness. Because the irrational numbers are infinite. They can't be written down. They force human consciousness out beyond the limits. And by adding irrational numbers to rational numbers, you get real numbers. . . . It doesn't stop. It never stops. Because now, on the spot, we expand the real numbers with imaginary square roots of negative numbers. These are numbers we can't picture, numbers that normal human consciousness cannot comprehend. And when we add the imaginary numbers to the real numbers, we have the complex number system. The first number system in which it's possible to explain satisfactorily the crystal formation of ice. It's like a vast open landscape. The horizons. You head toward them and they keep receding.

PETER HØEG, *Smilla's Sense of Snow*

Contents

A Note on Taxonomy

As the title of this field guide indicates, the various genera of numbers can be considered as forming a kingdom, not unlike those that classify the species of the living world. This realm might be labeled something like Kingdom Number. Further, I have adapted the binomial Linnaean system from biology to mathematics, giving every number a genus name and a species name. Instead of, for example, the genus *Homo* or the genus *Rosa*, there are such genera as *Natural* or *Real*; instead of a species such as *sapiens* or *rugosa*, there are species such as 7 or π.

There have to be some adjustments, however. Although each number, like each species, is unique, the genera in mathematics appear to overlap. Thus, the number 5 is commonly seen behaving as a counting number with the binomial name *Natural* 5. But 5 may also be found in a flock of fractions, where it is clearly *Rational* 5. Sometimes it is helpful to write *Rational* 5 as 5/1 to show that in a given case it is indeed *Rational* 5 that is being considered and not *Natural* 5. Similarly, since 15 and 5 behave alike, it is easy to treat *Integral* 5 as if it were *Natural* 5. In the operations of ordinary algebra or calculus the 5 one sees is *Real* 5, but in more advanced mathematics or certain engineering contexts the number 5 is always *Complex* 5, whether or not it is written as $5 + 0i$. In all these instances, the genera (their defining characteristics to be found in the following pages) only appear to overlap: for nearly all purposes you can use *Natural* 5 and it will behave just like the 5 from any of the other genera.

There are other useful number classifications that do not rise to the level of genera. These others, which do not encompass an entire system of numbers with attendant operations, I call families. Two examples of very large families are the Irrational Family and the Transcendental Family, each a family of real numbers. Among the natural numbers there are families such as the Odd Family or the Prime Family. One way to recognize that a family is less than a genus is to watch the behavior of the members for addition and multiplication. Notice that for the genera *Natural, Integral, Rational, Real,* and *Complex,* the sum or product of any two members of a genus is within that genus. But for families this condition, called closure, fails. It is easy to find pairs of irrational numbers for which the sum or the product is not irrational: think of $(+\sqrt{2}) + (-\sqrt{2}) = 0$ or $(+\sqrt{2}) \times (-\sqrt{2}) = -4$, for example. Similarly, the sum of any two odd numbers always fails to be odd and the sum of any two prime numbers may or may not be prime, while the product can never be prime. Although multiplication in the Odd Family shows closure, that is not enough for genus status. Indeed, there are some families, such as the Even Family, that show closure for both addition and multiplication, but by common consent are not considered important enough to be call genera.

While this system is useful as described, complicating it further with orders, classes, and phyla leads only to a messy situation (the thoughtful person might note that this is also true of biology, but biologists and especially botanists seem not to care).

Thus the various species and genera of numbers form Kingdom Number. The various forms of infinity—which are not themselves numbers—belong to a different kingdom, which I call Kingdom Infinity. Auden's phrase "the Kingdom of Infinite Number" captures both these concepts.

In this book, the two kingdoms are presented separately. The emphasis and most of the space are devoted to Kingdom Number, which comes first. Each genus is discussed in general terms, then representative samples—"species"—of that genus are presented in order of size. If a number, such as 5, occurs in more than one genus, appearing as it does as *Integral 5*, *Rational 5*, *Real 5*, and *Complex 5*, the species that is least inclusive is the one treated—so, for example, -1 appears as *Integral* -1, 1/5 is treated as *Rational 1/5*, and $\sqrt{2}$ is discussed as *Real* $\sqrt{2}$, not as *Complex* $\sqrt{2}$. In the discussions of individual species, *Field Marks* describe how to recognize the species, and *Similar Species* are briefly introduced. In *Personality*, you'll see how knowing the inner workings of the particular species aids in computation with that number. Finally, *Associations* points you to places you might encounter the species, and suggests why it can be found there. Along the way, boxes detail interesting families of numbers and occasionally other topics related to numbers. Frequent cross references, keyed to the running heads, help to keep all this straight.

Acknowledgments

I would like to thank the staff at W. H. Freeman and Company for their help in producing this book, especially my editor John Michel, who made significant contributions to the form of the book; Nancy Brooks, whose copyediting transcended copyediting; and Jane O'Neill, who shepherded a difficult production task and schedule. I also want to thank my wife, Mary, for her contributions as first reader and aid in indexing and for overseeing the author.

Introduction

The intent of this field guide is to aid the reader in identifying numbers in their native habitats. Just as an experienced birder can tell which birds are hidden in the trees by hearing their songs, or recognize a species from a silhouette or a flash of color through the leaves, the experienced number-watcher learns to find the hidden secrets of numbers, to classify a number instantly, and to use number relationships to enhance the enjoyment of mathematics as well as to solve problems. Just as nearly everyone can recognize a robin on the lawn or crow's caw from the sound, most adults can easily spot one of the smaller perfect squares, such as 25 or 49. But the adept birder identifies which warbler is singing in the woods and knows the profiles of all the birds of prey. Similarly, better number-watchers know that 10 is a triangular number, that 21 is in the Fibonacci sequence, and that π is transcendental. The very best catch relationships such as the "amicable" one between 1184 and 1210.

Often becoming a good number-watcher is just a matter of awareness. People simply fail to notice the numbers that flit though the forest of problems that daily surrounds us. I often begin a new semester of teaching math by asking the class to name numbers, challenging each student to propose a number *as different as possible* from the ones that have come before. I chalk each number on the board as it is put forth and discuss its particular properties—is this one larger or smaller than the others, is the number odd or even, prime or composite, positive or negative? Nearly always the first few numbers students name are small counting numbers,

numbers such as 7, 23, and 4. As I press the students to be more original, someone will finally think of a fraction or a decimal, someone else of a negative number. In better or larger classes, irrational numbers such as $\sqrt{2}$, and perhaps even such transcendental numbers as π or imaginary numbers such as i, will finally emerge.

Numbers and the Mind

We are born with a concept of number. Many number ideas, and often even the formal number classifications, are present in the human mind, but not, apparently, where they can easily be called forth. Practice and awareness bring number ideas to the front of the mind and unmuddle them. One main purpose of this field guide is to list the properties that make a given number unique and that are essential in mental computation. For the most part, these properties are not obscure mathematical concepts (although sometimes a number has such a special quality that it is worth going into a bit of mathematics to explain it). Instead, they are simple properties, such as the way the number can be formed by multiplication or addition. It is also worth noting what general type of number is under discussion: some numbers are famous for a particular property. Just as dice players are likely to recognize 7 as lucky (it's not just a superstition), mathematicians may view 6 as perfect and immediately think of 8 as the cube of 2.

Knowledge of the personal traits of the numbers is often useful in computation. For example, consider the following problem, the final question in the 1997 U.S. national Mathcounts competition: "Shelly's total cost for a mountain bike is $900, including tax and interest. Her first payment is 25 percent of the cost and she pays the remainder of the cost in 15 equal monthly payments. How many dollars is one monthly payment?"

The thirteen-year-old winner reasoned as follows: "She pays $900 total. A quarter payment is $225. Subtract that from $900 and you get $675. There are 15 equal payments so you divide by 15." But that is not the way I saw the problem, which I solved quickly before reading the winner's analysis. I immediately thought of 25% as one fourth, which is half of 1/2. Using this, half of $900 is $450 and half of that is $225. But since I recognized 225 as the square of 15, I saw that 15 monthly payments would mean $15 each. Since this amounts to only 1/4 the total, and the remaining 3/4 must be paid, the answer is 3 × 15, or $45 for each payment. We both got the correct answer, but I could do all the calculations easily in my head. In fact, I don't think I could have used the traditional way without resorting to paper and pencil. Indeed, I could not have subtracted $225 from $900 in my head and reached the correct difference for that part of the computation in the traditional way as fast as it took me to think through the whole problem. Furthermore, it would not have been easy for me to have divided $675 by 15 to get $45 if I were using the ordinary division algorithm.

What is it that enabled me to make such a quick calculation? It all stems from familiarity with the personalities of specific numbers. Although thinking of 25% as 1/4 is not all that unusual, nor is seeing 1/4 as 1/2 of 1/2, I find that my freshman college students often fail to recognize these relationships.

But the key to my calculation was the quick recognition of 225 as the square of 15. When I was about twelve years old, I noticed that I needed the squares of the whole numbers fairly often. I had learned the times tables through 12 × 12 in third grade, as all students were once required to do. So I added to my list by learning that 13 × 13 is 169, 14 × 14 is 196, and 15 × 15 is 225. For good measure I also learned 25 × 25 is 625. These perfect squares have been helpful companions ever since.

The evidence is that these ways of thinking can be strengthened in anyone, producing not only better computation, but also a better understanding of the world that comes from recognizing the patterns of the universe, which somewhat mysteriously seems to have been written by the Creator in the language of mathematics.

Studies of infants and children, whose brains are not fully developed, and of people with damaged brains have revealed much about the way that we process numbers. Perhaps the most surprising discovery is that humans (and probably some other animals) are born with an innate number sense that appears to be in the form of a number line:

This number line can be observed at different scales. When you zoom in on it mentally, so that the scale is very great, you perceive only the first few of the positive numbers, but these are easily seen. With this mental tool, hardwired into the brain, you can easily recognize the numbers 1, 2, and perhaps 3. Although the numbers are perceived along the number line, the brain seems to have no difficulty in connecting these measurements to discrete objects immediately without counting them, recognizing, say, the difference between two marbles and three. Not only are the very small counting numbers perceived directly, but certain operations with them are also innate or perhaps learned extremely early. Most people use this mental number line to determine, for example, that 2 is between 1 and 3 and that 3 follows 1 and 2, rather than needing to remember and apply these statements in each case.

Babies as young as five months can be tested on their ability to perceive and operate with numbers because a baby

4

will spend more time looking at something unexpected than at something expected. If the result is an expected one, the baby quickly looks elsewhere for novelty. If the result is unexpected, the baby's rather short attention span is held a bit longer.

Using this premise, investigators measured infants on their ability to tell whether $1 + 1 = 2$ and whether $2 - 1 = 1$. A doll was hidden behind a screen in view of the baby. Then a second doll was put behind the screen. Sometimes when the screen was lifted, the two dolls were both there. But sometimes one of the dolls was slipped into a hidden compartment, so when the screen lifted the baby saw only a single doll—and studied the scene much longer. A similar experiment showed that the baby could perceive that $2 - 1 = 1$ is true and that $2 - 1 = 2$ is incorrect. Interestingly, studies of adult monkeys using a similar technique show that they possess the same mathematical abilities as infant humans do. Monkeys have also been shown to be able to place small sets of objects in order of number, effectively counting, but counting of this type has not been demonstrated in infants.

The experiments with infants could by themselves be interpreted as showing either a built-in number line or an inborn sense for discrete small numbers. Studies of older children begin to discriminate between the two possibilities. A series of famous experiments by Jean Piaget demonstrated, among other things, that young children do not recognize that number is conserved—that is, that the same five objects clustered together are the same in number as five objects spread apart. In light of other studies, this finding can be interpreted as follows: young children use their innate number line to measure the number of objects. On the number line, magnitude (size) rather than correspondence (matching with a mental set) is the actual measurement; the mind then has to convert the size of the object into discrete units. When the object has been expanded (by spreading apart the units), the child perceives that the number of units has been increased. Older children learn, by spreading objects out or by pushing them together, as a separate piece of information that number is always conserved, so they override the signal they are getting from their innate number line.

The proof of this interpretation comes from studies of people whose brains have been damaged by stroke or accident. There is a spot in the brain, on the left side for right-handed people, just above and behind where the ear attaches to the head, that contains, among other mental tools, the mental number line. If damage occurs at this location in the brain, the mental number line or perhaps some other aspects of number sense can vanish. If the mental number line is gone, the sense of simple operations may also be lost. Not only does the afflicted person cease to recognize that, say, 2 is between 1 and 3, but also the difference between odd and even disappears.

If the mental number line is still there, but somewhat fuzzy, the patient retains a vague sense of number. When asked about such familiar numbers as the number of months or days in a year, the patient may answer "fifteen months" or "about 350 days." The number line remains, so there is a general idea of size; that is, the patient does not say there are 350 months in a year or fifteen days in a year. Similarly, a patient asked the sum $2 + 2$ might reply "3" or "5," but not "30" or "200." Subtraction skills with small numbers also vanish when this region of the brain is out of commission.

Interestingly, studies of brain-damaged arithmetic show that the times tables are stored in a different part of the brain completely, in a region for memorized data. Thus a person who can no longer add $2 + 3$ because of brain damage can usually multiply 2×3 and even solve more difficult

6

multiplication problems. Errors in remembering, however, cannot be checked against the mental number line. A person who somehow remembers 6 × 8 as 84 instead of 48 and has lost the mental number line (or fails to use it) will not notice the error.

Since the number line can be zoomed in close for small numbers, operations with these can be computed exactly even when other number information is lost. But to use the number line for large numbers one has to zoom out. Thus, it is more difficult to determine whether or not an error has been made in calculations involving large numbers. The main techniques for handling arithmetic are all based on replacing large numbers in operations with small ones. Even if you need to add 102 + 3 you mentally ignore the 100 in 102 while you are doing the operation. If you need to estimate 537 + 281 you may add 5 + 2 or, better, 5 + 3 and then think "hundreds" to get estimates of 700 or 800. Similarly, you estimate the product 537 × 281 by finding either 5 × 2 or 5 × 3. It probably does not matter to your brain that the addition problems are largely handled in the number-line region while the multiplication is located over by the memorized dirty limericks.

Auden's lines, "Instead of a nodding acquaintance with a few integers, / The Transcendentals are our personal friends," are a glancing reference to the self-taught mathematical prodigy Srinivasa Ramanujan, of whom it was said, "Each of the positive integers was one of his personal friends." [This remark is usually attributed to the Cambridge mathematician John E. Littlewood, but according to Howard Eves, writing in *Introduction to the History of Mathematics*, G. H. Hardy used it in an article, introducing it "As someone said. . . ." Littlewood, who proofed the article, commented, "I wonder who said that; I wish I had." And somehow, by the time the article was printed, the sentence had become, "It was Littlewood who said. . . ."] The greatest mathematicians are known for this ability to recognize hidden number patterns as we recognize the faces and traits of those we know well. Karl Friederick Gauss, born near the end of the eighteenth century and still thought by most historians to have been the best of the lot since the invention of mathematics, was endowed from an early age with this ability to recognize patterns and improved it by practice during his long life. The well-known story of how as a child he summed the numbers from 1 to 100 illustrates the kind of recognition of friendly numbers that this book is intended to encourage. His teacher assigned his class the task of adding all the numbers from 1 through 100 as busy work. Gauss did not attack this problem at all as the teacher expected, which was to say 1 + 2 is 3, 3 + 3 is 6, 6 + 4 is 10, and continue in that way to 100. Instead, Gauss began by thinking 1 + 100 is 101. Also, 2 + 99 is 101, the same sum. This pattern will continue with 3 + 98 = 101 and so forth up to 50 + 51 = 101. There will then be fifty of these friendly 101s to add, so the problem is really 50 times 101. [At that point I would think that multiplying by 50 is putting two 0s at the end (to multiply by 100) and dividing by 2, so the answer is half 10,100 or 5050; but most histories of math suggest that Gauss multiplied 50 × 101 to get the 5050, which is actually not difficult to do.] He wrote down the answer, 5050, almost immediately, to the great annoyance of his teacher, who was still adding 10 + 5 is 15 and 15 + 6 is 21 and so had no way of knowing whether Gauss was right or wrong.

The key to learning how to compute mentally ("in your head" instead of on paper with a pencil) is recognition of which numbers can be treated in specific ways to simplify the computation. If you become adept at this, you recognize that numbers close to multiples of 10 (1 or 2 numbers away) must be handled differently from numbers farther away. You learn to spot multiples of 3 or of 8 quickly and to exploit the special properties of such friendly numbers. You learn how to compute an answer one way and then use a completely unrelated approach to check your arithmetic.

In mental computation, especially multiplying or dividing, you can replace unfriendly numbers in a problem with those that are kinder and gentler. The most common way to make such a replacement for multiplication or division is to pick and choose among the factors. If you have a good relationship with the numbers 35 and 16, for example, you instantly think of them as $35 = 5 \times 7$ and $16 = 2^4 = 2 \times 2 \times 2 \times 2$. Thus, the product 35×16 is the same as $5 \times 7 \times 2 \times 2 \times 2 \times 2$. Now you pick and choose to make your best deal. For example, you can pick out the 5 and one of the 2s to make $5 \times 2 = 10$. Then what is left is a 7 and $2 \times 2 \times 2 = 8$ from 16. Combine the leftovers as $7 \times 8 = 56$. You have replaced the moderately difficult problem 35×16 with the very easy problem 10×56 that has the same product. So the answer is $10 \times 56 = 560$.

Or you could have seen the same problem of 35×16 in a different light and collected the 2, 5, and 7 from the factors, to make 70. What is left are the three 2s, or 8. Then the problem is 70×8, which is also easy to recognize as 560.

There is no single best way to compute with friendly numbers; it depends on which ones are comfortable for you. When I learned the multiplication table in the third grade, for some reason the most difficult product for me was 7×8 (others also find this one difficult). I still take a fraction of a second longer to remember that 7×8 is 56 than most other multiplication facts. So it might even be easier for me to avoid that difficult situation and mentally compute 35×16 in a different way entirely, one that avoids 7×8. Here is that other approach.

Since 16 is 2^4, or $2 \times 2 \times 2 \times 2$, I might turn to one of my favorite computational techniques, doubling, which in several societies is the main method for multiplying any pair of numbers. I would think double 35 to get 70, double again is 140, double again for 280, and the answer is twice 280 or 560. Same answer, of course, but I found 16 to be a very friendly number, while 7 and 8, taken together at least, are less friendly.

Numbers and Writing

Once every eight days, I fill a small weekly pill box with the vitamins, minerals, and medicines that according to various theories will help me live longer. There are seven compartments to fill, one for each day of the week ahead, and I also put aside pills to take today. As I shake out each group of pills onto the tablecloth, I look for exactly eight. Eight is too big to recognize. It is easy to see two fours or even a five and three, however, and in this operation, all I need is number sense and a steady hand. But for nearly every operation with numbers greater than 10 or 12 I need to think of them as numerals—as they are written in our number system.

Sometimes humans think in pictures or concepts without words, but most psychologists believe that thought usually involves language.

Just as most of us think in a particular verbal language most of the time, we confront numbers using the ways developed for writing those numbers. We are taught to think in the Hindu-Arabic decimal system of numeration, not in Roman numerals or even with tallies. If we had learned to think in Roman numerals or with tallies we would compute in a different way. We would perceive numbers in a different way—at least larger numbers. Small counting numbers are "given by God"; they are hardwired into the brain. But larger counting numbers are not—we have to use a language to think about them, and the language that we are taught determines how we think.

Methods of writing numbers are called *numeration systems* if they are systematic structures, although there are also methods of writing numbers so closely associated with numbers that they are almost pictures of numbers instead of numeration systems. For example, we often keep score of games with the short strokes called tally marks. Usually these begin as simple strokes, I II III IIII, and they could continue that way if scores in the game are low: IIIIIIIII to IIIIIIII, for example. But it is hard for the eyes to track very many of these, and if scores run high, people fall into using a simple numeration system of tally marks, striking off every group of five objects or events by making the fifth tally horizontal instead of oblique—so ⅢⅢ means 5. Numeration systems strongly affect the way we compute, just as language affects the poems we write. To understand computation at all, one needs to know the basics of numeration.

The history of computation begins with devices, not with the mind. The original methods of computing precede writing; indeed, the evidence strongly suggests that in the early days of civilization, in the Middle East, writing emerged as an outgrowth of the recording of computation.

Before the Neolithic Revolution, when agriculture became a serious business and trade and vocational specialization became entwined with urbanization, there was not much worth counting, although there is some evidence that counting occurred. Certain ancient bits of bone or ivory dating from as early as 30,000 years ago have been found with what appear to be tally marks carved on them. These marks are grouped in such a way as to suggest that the most likely entities being counted were the days, or rather the nights—phases of the moon appear to be involved. An astronomer can choose to look at these day counts as an early calendar, but a mathematician must see numbers in the scratches on bone.

In cave paintings from about 10,000 to 15,000 years ago, one can more clearly see this idea of number as tallies. A wounded animal is shown once, but four ocher slashes are painted next to it. We infer four successful hunts.

The next evidence for counting would, by itself, be impossible to interpret. It is only by knowing what follows this evidence and by working backward that archaeologists, especially Denise Schmandt-Besserat of the University of Texas at Austin, have recognized that the small baked clay pieces now known as tokens were used as counters. Early in the 1970s Schmandt-Besserat began a study of the use of clay by humans before the invention of pottery. Consequently, she began to evalute any ancient objects made from clay that were found in museum collections or at archaeological sites, especially in the Middle East, where the pottery wheel was invented and where fired bricks first came to be used in building. She looked at everything, from clay floors in houses to clay beads used as ornaments, and discovered that archaeologists had found thousands of small clay geometric shapes—among them cones, spheres, tetrahedrons—whose purpose was unknown. These small clay objects had been collected and cataloged by archaeologists since at least 1905, but no one had a clue what they might be.

Some other clay objects had been unearthed, also mysterious, but clearly connected to numbers. One such, found in 1920, was an empty clay holder, or envelope, that was labeled on the outside in cuneiform writing as containing "counters" that represented numbers of different kinds of sheep and goats, 21 ewes, 8 rams, and so forth, totaling 49 animals. But the forty-nine counters originally inside the envelope had been lost. In the 1960s French archaeologists studied some much earlier clay containers that held hollow clay spheres. These predated the development of cuneiform writing, but the clay spheres had odd geometric markings on their surfaces. The French archaeologists Pierre Amiet and Maurice Lambert recognized that the markings on the outside of the containers were reverse images made by pressing the spheres into the soft clay of the containers as they dried.

A few years later, when Schmandt-Besserat was seeking meaning in the large number of geometric clay objects that had been found over the years, many of them from sites even older than those that produced the clay spheres of Amiet and Lambert, she read Amiet's article on the spheres and had the moment of insight that was to shape her work from then on and that would change entirely our understanding of the development of writing. The small clay geometric objects were the earliest version of the counters that had been originally in the clay envelope from 1920, she reasoned. She chose to call the objects "tokens" to suggest that they were used in trade to mean numbers of objects being traded. The geometric shapes were simply identifiable forms that are easily made by hand from bits of clay. A particular shape for a token could then be assigned a meaning—a cone might mean a sheep, a sphere a goat.

On the basis of peasant practices throughout the world, histories of mathematics have frequently suggested that before numbers were known farm animals were counted by one-to-one correspondence. A farmer who had twenty-seven sheep might keep a container of twenty-seven pebbles, one pebble for each sheep. As the sheep were brought into the fold for the night, the stones could be moved from the container to another spot. If all the stones were moved, all the sheep were in the fold. If there was a stone left in the container, a sheep was missing and should be sought. When sheep were bought, sold, born, or died, the account could be kept up to date by adding or subtracting pebbles. Although some form of this pre-numerical system is commonly put forth, such accounts have largely been speculation by the writers, with little evidence of actual practice. But Schmandt-Besserat realized that this basic idea of a pre-numerical concept of number is correct, and also that such an idea of number began with the practices of early farmers and traders some 10,000 years ago.

The clay tokens were used in various forms for about 5000 years, from the dawn of the Neolithic Revolution until well into the early urban civilizations. During that lengthy period, as long an interval before the beginning of history as time has passed since, the concept was reworked and improved. Standard forms of tokens were developed, and the practice of keeping tokens in hollow clay spheres began. It is thought that these spheres were used in long-distance trade. A servant or partner could be entrusted with the sheep or goats (or jars of oil or wine or wheat or bars of metal) and made to carry to the customer a hollow clay sphere with tokens inside that represented the cargo. Upon its arrival, the customer could break open the sphere and count to make sure that everything promised had been actually delivered. This interpretation is back in the realm of speculation, of course, but the evidence is strongly suggestive. Let us continue to speculate.

Someone then had the idea of marking the outside of the spheres with impressions of the tokens inside; marked spheres do not have to be broken to check the contents. This practice apparently continued for several hundred years before people observed that if the message was on the outside of the sphere, there was no need to put counters inside the sphere. Indeed, there was no need to go to the trouble of making a hollow receptacle for tokens. Thus, people began to use small rectangles of clay, conventionally called tablets, to record numbers. Different counters could be marked right on the face of the clay tablet.

People eventually learned that it is easier to use a single marker and press it into the clay in several configurations than it is to have different tokens that each have to be pressed. Perhaps in the beginning there were a sphere, a pyramid, a cube; but a better idea was to have one object that could make several different symbols. A short piece of reed, round at one end and wedge-shaped at the other, could produce three different marks that could be combined to create an infinite number of different symbols. The marks that such a reed stylus could make in clay became the basis of a number writing system.

The early numbers are thought to have been all denominate—that is, one pair of marks meant two sheep, while a different pair of marks meant two jars of oil. We have remnants of such a system in English: we speak of a "brace of quail" but "a pair of dice." The reed stylus was used originally to record different numbers of objects by pressing its mark in the clay as many times as the number. If, for example, the symbol ● meant a jar of oil and ◀ meant a lamb, then ● ● would mean two jars of oil and ◀◀ would mean two lambs.

Because the first numbers were denominate, the line between writing numbers and writing words is somewhat blurred in the first stages of written language. When numbers were freed from specific objects and made abstract, writing was invented. Say that a new mark is invented that always means 2, no matter what the object. One such early mark was a pair of parallel lines, rather like our equals sign, =. With the new mark for 2, then = ◀ becomes the symbol for two lambs while = ● is the symbol for two jars of oil. In this view, writing words was an outgrowth of the system for writing numbers.

From the point of view of computation, many of the most useful properties of numbers are those based in the system used for writing them. For example, the methods used for recognizing which numbers can be divided evenly by 2, 3, 4, 5, 6, 8, 9, or 10, a fundamental part of mental computation, are all based on the way numbers are written and not on the numbers themselves. This distinction between the written symbol, or *numeral*, and the number itself is of fundamental importance to logicians and even to some mathematicians, although it seems so trivial to most people that it was among the gravediggers for the "new math" of the 1960s. Mathematician and logician Charles Dodgson, writing as Lewis Carroll, made fun of it in the *Alice* books, notably when the White Knight (often viewed as a surrogate for the author) attempts to cheer Alice with a song:

> "You are sad," the Knight said, in an anxious tone: "let me sing you a song to comfort you. . . . The name of the song is called 'Haddocks' Eyes.' "
>
> "Oh, that's the name of the song, is it?" Alice said, trying to feel interested.
>
> "No, you don't understand," the Knight said, looking a little vexed. "That's what the name is called. The name really is 'The Aged Aged Man.' "
>
> "Then I ought to have said 'That's what the song is called'?" Alice corrected herself.
>
> "No, you oughtn't: that's quite another thing! The song is called 'Ways and Means': but that's only what it's called, you know!"
>
> "Well, what is the song, then?" said Alice, who was by this time completely bewildered.
>
> "I was coming to that," the Knight said. "The song really *is* 'A-Sitting on a Gate': and the tune's my own invention."

While it sometimes seems as if the mathematical distinction between the number and symbol for it is as batty as the

White Knight's over-careful labeling of his song, the idea can really be helpful once a person becomes accustomed to it. From the symbol point of view, for example, it is helpful to think of 7 + 3 as just another name for the number ten, which I write here as t-e-n to avoid the symbol 10 that means specifically 1 ten and 0 ones, which would hardly make sense as 1 10 and 0 1s—or would it? Perhaps I am as batty as the White Knight. In any case, it is sometimes very helpful to be careful about making the distinction between a name and the object named, but most of the time it is unnecessary and rather foolish. I try to maintain some balance in what follows.

The modern system for writing numbers was compounded from various sources. It is known as Hindu-Arabic because its most obvious, and most recent, genealogy is the typefaces that were designed to match the handwritten system of numeration used by the Arabs and other Islamic peoples at the time of the Crusades, the twelfth and thirteenth centuries. Scholars have established that this system was an Arab modification of the numeration used by people living in India about A.D. 1000 and two or three centuries before that, before the Islamic occupation of much of the subcontinent. As most of those Indians practiced Hinduism, the name Hindu-Arabic is commonly used today. But that system resembles various other numeration practices of Asia that date back much before this time. If one looks at the Indian numerals, or signs for numbers, from the first millennium, it is easy to see a cousinship with Chinese numerals from around the same time and earlier.

Indeed, all systems of numeration seem to have evolved from several primitive practices whose remnants are clearly visible in the signs and the underlying systems that persist today. The most basic form of indicating numbers is with body parts. Counting using fingers or other parts of the body is common in societies without writing and was widespread in Europe during the Middle Ages, used by peasants and literate scholars alike.

It is easy to see, for example, that 5—as in the five fingers on a hand—has been an important stopping place for numeration. This is clearly visible in Roman numerals, which use V for 5 and two Vs in the form of an X for 10. The Maya, whose numeration system probably was developed independently from the Asian-based systems, used dots for 1, 2, 3, and 4 and then a bar for 5. If you look closely at the Hindu-Arabic numerals for the first five numbers, you can see that they are constructed from strokes that are connected to make for speedier writing. 1 is just the single stroke, written as a vertical segment in the tradition of Sumerian and Egyptian numeration. 2 is two horizontal strokes connected by the scribe's keeping the pen on the parchment, in other words essentially a Z. (Horizontal strokes for the beginning numerals are a feature of East Asian numeration, especially the Chinese system.) 3 is similarly three connected horizontal strokes. 4 as written or printed appears also to have originated as three strokes, but the earlier forms are more clearly formed by combining two horizontal and two vertical strokes. 5 is also produced from a combination of horizontal segments, in this case three, and two vertical strokes. After 5, however, a different system seems to come into play. The numerals 6, 7, and 9 are difficult to construct from that number of strokes, although at a stretch it is possible to see eight strokes in 8.

Similarly, 10 is both the fingers on two hands and the main stopping place for many numeration systems. This number is used as the main basis in the old Egyptian hieroglyphic system and in the Chinese system and provides a subsidiary basis in the most ancient Mesopotamian system. It is the foundation of the Hindu-Arabic system as well. Thus, another name for the Hindu-Arabic system is the decimal system, *decimal* from the Latin for "ten."

There are several types of decimal system, however. Most of the early ones were simply additive. Mentally, this process is like grouping sets of objects in bundles of ten and then counting the number of bundles. Thus, the Egyptian system used a stroke, I, for a one and a hoop, ∩, for ten. To write a number such as forty-seven they simply put down four hoops and seven strokes: ∩∩∩∩IIIIIII. The order and grouping were not important, although the hoops were usually put all in one place and the strokes together in another.

The earlier Roman system worked much like the Egyptian, with an additional symbol for five. The numerals for one to twenty looked like this:

I II III IIII V VI VII VIII VIIII X XI XII XIII XIIII XV XVI XVII XVIII XVIIII XX

Later Roman scholars added a subtractive feature to go with the basic additive format. If the numeral for a smaller number is written *before* a numeral for a larger number, then the smaller number is subtracted from the larger one. With this change Roman numerals from one to twenty assumed the more familiar form that is used on some clocks or occasionally in dates, especially dates that one wishes to obscure (such as perhaps the copyright date on a motion picture):

I II III IV V VI VII VIII IX X XI XII XIII XIV XV XVI XVII XVIII XIX XX

The Hindu-Arabic system represents a much better solution to the problem of writing larger numbers. Instead of combining addition and limited subtraction, it combines addition and multiplication with a complete system based on 10, a system in which the place of a digit in a numeral indicates its value to the number—thus, a place-value system. The fundamental simplicity and regularity of the Hindu-Arabic system has been the reason that this system, or variations on it, has by now replaced all other ways of writing numbers (except, ironically, in Arabic, in which symbols continued to evolve after the European borrowing, so today's Arabic symbols are different from those used by nearly everyone else on Earth). The Hindu-Arabic system is so pervasive that most people around the world equate the way numbers are written in the Hindu-Arabic system with the numbers themselves, failing to separate the properties of the number from the properties of the way it is written.

And so, let us look at the numbers themselves and also use, when we can, the ways we represent them. As we travel through the number species it is good to remember that every one of them is interesting and even exciting. For numbers that can be arranged by size, this can be proved. Suppose that some numbers are not interesting. The set of such boring numbers can, as agreed, be arranged by size, so one of the dull numbers is the smallest member of the set. But that makes it a number of interest, since it is the smallest number known that has no unusual properties. This contradicts the premise that all the numbers in this set are without interest, so such a set must not exist. All numbers are interesting.

Kingdom Number

Genus *Natural* (Counting Numbers)

In some societies, counting language has not evolved beyond 2 or 3 and a word that means "many" for higher numbers. From a linguistic point of view the counting numbers, at least those greater than 2 or 3, may be considered a human invention.

Yet there is something natural about the genus *Natural*. If the concept of matching, or one-to-one correspondence, is taken as a fundamental idea, the natural numbers— 1, 2, 3, . . . —would appear to be its logical outgrowth; a natural number describes the set of all sets that can be matched a member at a time with a given set, without regard to order.

From this point of view, 1 might be considered the characteristic that all the sets which match with me myself, considered as an individual, have in common. The number 2 is characteristic of collections that match my eyes or hands or legs or ears. Then 3 is part of the description of sets that match the spaces between the knuckles on one hand, 4 the sets that match the fingers without the thumb, 5 with the thumb, and so forth. You see why some societies stopped using natural numbers after 2 or 3, since we quickly begin to exhaust the observables shared by all.

Although the idea of one-to-one correspondence may seem natural—it is pretty much what we teach to young children as the meaning of number—it is not very mathematical as most mathematicians for the past 200 years have envisioned the field. Mathematical ideas must be pinned down; it is not enough to say that we abstract a concept from experience. The definition of a mathematical entity must be "operational"—that is, there must be some operation that produces the entity. Furthermore, the operation needs to emerge from within mathematics. The ancient Greeks were satisfied, for example, with the idea that geometric figures are congruent if one may be picked up and superimposed on another. This intuitive idea is rejected today because superimposition is not a mathematical operation. Congruence is established instead with a series of definitions based on such ideas as equality of lengths or angles, which are part of mathematics generally. So it is with the numbers—instead of the "natural" interpretation of numbers, mathematicians use the "counting" interpretation. The nineteenth-century German mathematician Richard Dedekind, the first to define the genus *Real* without recourse to geometry, wrote: "I regard the whole of arithmetic as a necessary, or at least natural, consequence of the simplest arithmetic act, that of counting."

Several mathematical definitions based on counting have been proposed for genus *Natural*. The most common and in some ways the most acceptable was developed by the Italian mathematician Giuseppe Peano [1858–1932] in 1889. It combines the idea of counting with a very general principle called mathematical induction that applies nicely to counting numbers. The Peano definition is a set of five axioms that begins simply and becomes increasingly less intuitive, but still manages to have the axiomatic property of seeming inevitable (although a certain amount of reflection is required to grasp the essence of the last two). Those who are familiar with Euclid's five postulates for geometry will notice a certain parallel.

In ordinary language, Peano's definition is as follows:

The number 1 exists.

You can always count further by adding 1 more.

No amount of counting will bring you back to 1 again.

If two numbers reached by counting are equal, then the numbers just before them are also equal.

If you define a set of numbers so that: (a) *if* the defining property applies to some counting number, *then* it also applies to the number that is 1 more; and (b) *if* 1 has the property, *then* all the counting numbers also have the property.

The last rule in the definition is hard to follow, with its two *ifs* and two *thens*, so I have tried to emphasize the parts. Peano himself would never have stated these rules in anything like this language, however. He believed strongly in symbolic logic and stated all his conclusions using special symbols, so his five-volume book explaining why this definition describes the counting numbers is almost all symbols and unreadable in the ordinary sense. He also insisted on writing out all his lectures using symbolic logic, which caused the students in his classes to rebel. Peano tried to quell the rebellion by promising the students that they would all pass, no matter what, but this attempt at pacification failed. As a result, Peano was forced to resign from teaching.

The terms "counting number" and "natural number" refer to the same set of numbers, but emphasize different aspects of them. The genus name for these numbers is *Natural* because that is the name mathematicians usually use, but the definition that formal theory prefers is based on counting rather than the "natural" idea of matching sets. Thus, it would be more logical to speak of the "counting numbers" rather than the "natural numbers." In some formal theories (for example, the conditions of the famous proof known as Gödel's Theorem—*see* **Prime Family,** pp. 35–38) it is important to make a clear distinction between the natural numbers and the counting numbers, but that is too fancy for this project. I generally use whichever term fits the context.

Commonly Seen Species

1 •

1 One I unity

French *une*; **German** *ein*; **Spanish** *unos*; **Italian** *uno*; **Latin** *unus*; **Greek** *heîs*; **Papuan body counting,** right little finger; **classical and medieval finger numerals,** left little finger folded at joints, thumb raised

There is only *the* one, but it comes in two guises. *One* is sometimes a number, but sometimes not; sometimes it is a discriminator.

The word "discriminator" extends the grammatical idea of the article. Articles form the smallest grammatical category in English, comprising only two members: "a" and "the." The main function of an article is to indicate that what follows immediately is a noun and not some other part of speech. This function is sufficiently unnecessary that part of it is dropped in other languages such as Russian, which has no word for "the." Thus: *word "discriminator" extends grammatical idea of article*.

The broader class of discriminator includes, along with the articles, such words as "this," "that," "such," and, of course, "one." "Pointer" would be a more useful class distinction than "discriminator," since what most discriminators do is point: this child, that elephant, such apples. The other discriminators are those that specifically *fail* to point: a child, an elephant, one apple.

"One" can also be a nonpointing discriminator, functioning approximately as a more emphatic "a" or "an." The sentences "There'll come a day when I'll find you" and "There'll come one day when I'll find you" mean essentially the same, but "one day" is stronger than "a day." This discriminator function of "one" gradually merges into the number function. If you say "Give me a rose" you mean the same as "Give me one rose," but in the second sentence the discriminator has drifted further toward numberhood. In the other direction, when used as a pronoun "one" has another nonnumerical use, which is distinctly unlike the number function. When I say "One starts counting with the number one," the first "one" does not mean a single individual, but quite the opposite. As a pronoun, "one" means "everyone"—the ultimate nonpointer.

FIELD MARKS The number 1 is the least counting number, the place where you begin to count. My older son, when still a preschool child, once argued with my attempt to teach him what the counting numbers are. I said 0 is not a counting number. He said, "Is too!" I said, "Show me. Count those chairs." He dramatically pointed his finger at the sky and said "Zero" and then pointed at the first chair and said "One." I was impressed, but not convinced.

The number 1 is also the number you "count by" in ordinary counting. That is, counting is ordinarily "by ones"—the equivalent of adding 1 each time you count. When a human "learns to count," he or she has learned the names of the numbers in order and matches each new item counted with the next number name from that memorized set: one, two, three, four, and so forth. But when a computer "learns to count," it does not use our language names. Instead, it adds 1 for each new item and assigns a name in machine language that is the sum "last number + 1." The human method is the reverse of the computer's: we consider that providing the next name is adding 1, while the computer adds 1 to obtain the next name.

The number 1 is the only number that does not change any number by multiplication or when used as a divisor. As such it is called the *identity element* for multiplication.

Other recognizable traits: 1 is the smallest triangular number (*see* **Triangular Family,** pp. 44–47), the smallest Fibonacci number and Lucas number (*see* **Fibonacci and Lucas Families,** pp. 120–122), the smallest square pyramidal number as well as a Catalan number (*see* discussion at **14** and **Square Pyramids,** p. 127), the smallest factorial (0! = 1!, *see* **Factorial Family,** pp. 145–146), and a Bell number (*see* discussion at **52**). Finally, 1 is the only counting number that is neither prime nor composite (*see* **Prime Family,** pp. 35–38).

SIMILAR SPECIES All other identity elements are modeled on 1. Among those commonly encountered are the identity element for addition, 0, and the identity function, $f(x) = x$. The identity element for addition is 0 because 0 does not change any number when added to or subtracted from any number. The identity properties of 1 and 0 are so nearly alike that when studying a mathematical structure it is difficult to tell what sets 0 apart from 1. The significant difference is that there is an operation that fails for 0—you can't divide by 0—but no such failure for 1.

PERSONALITY Since serially adding 1 is the same as counting, before learning the basic addition facts (the sums of all one-digit numbers), children say the name of the first number to be added and then count as many times as the second number. To add 5 to 2 you can say "two, three$_1$, four$_2$, five$_3$, six$_4$, seven$_5$" (the subscripts 1, 2, 3, 4, 5 keeping track of how many times you add 1). This technique is called "counting on," or, perhaps in honor of its digitally assisted version, "counting on your fingers." The virtue of "counting on" is that you have fewer ideas to keep track of while adding; you need think only of the number you are counting and how far you need to count for each addend.

The role of 1 in multiplication is to bring operations closer to tractable numbers. For example, multiplication by 100 is easy, as everyone knows. The friendly 100 can be combined with adding 1 or subtracting 1 to multiply by 101 or by 99. First, multiply by 100, then subtract or add the other number in the problem. For example, to find 99×37, think $100 \times 37 = 3700$ and $3700 - 37 = 3663$. This is a great deal easier than the ordinary multiplication method, writing the steps for "times 9," "times 90," and adding these "partial products."

This idea of using the two neighbors to friendly numbers greatly aids mental multiplication. For example, 25 is a friendly number because you can replace multiplication using a two-digit number with multiplication by 100 and division by 4. That means that the neighbors 24 and 26 are also easy. To multiply 26×54, for example, think $25 \times 54 = 5400 \div 4 = 1350$, then add 54 to get 1404.

ASSOCIATIONS   Part of the character and significance of each number comes from the associations it has in our minds and the uses to which it is put. Since 1 is associated with everything that is unique, there are more associations of this type than can be listed. In geometry, for example, any result that suggests a unique figure or point is associated with 1.

Examples include the angle bisectors of a triangle, as well as the altitudes, the perpendicular bisectors of the sides, and the line segments from a vertex to the middle of the opposite side (the medians), all of which meet at one point—not the same point, but one point only for each of the segments.

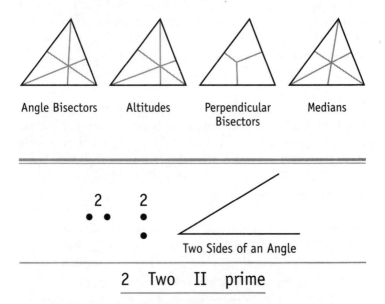

Angle Bisectors   Altitudes   Perpendicular Bisectors   Medians

Two Sides of an Angle

2 Two II prime

French *deux*; **Spanish** *dos*; **Latin** *duo* ("twice," *bis-* or *bi*); **Greek** *dýo* or *dís* (combining form, *di-*; "double," *diplóos*); **Old English** *tweegen*; **Papuan body counting,** right ring finger; **classical and medieval finger numbers,** left little and ring fingers folded at joints, thumb raised

Two separates the one and the many: nothing counts before you get to 2.

FIELD MARKS Recognizing twoness is easy. The same end digits repeat and repeat:

2	4	6	8	10
12	14	16	18	20
22	24	26	28	30 and so forth.

Once you learn the five digits 2, 4, 6, 8, 0, you can immediately recognize any multiple of 2. Is 25,678 a multiple of 2? Yes, it ends in 8, one of the magic five digits. How about 5,839,769? Nope.

The number 2 is the only even prime number and the smallest prime of all (*see* **Prime Family,** pp. 35–38). As the only even prime, 2 is also the smaller number in the only possible consecutive primes, 2 and 3. The prime 2 also figures in an obscure classification created by the French mathematician Sophie Germain [1776–1831]. A Sophie Germain prime is a prime number for which its double plus 1 is also prime; for 2, the qualification is met because $5 = 2 \times 2 + 1$ is prime. Although this classification starts out inclusively (the next two primes, 3 and 5, are also Sophie Germain primes), it then thins out. While Sophie Germain primes with thousands of digits are known, it is not known whether or not the number of such primes is infinite.

The number 2 is also a Fibonacci number (*see* **Fibonacci and Lucas Families,** pp. 120–122), a Catalan number (*see* discussion at **14**), a Bell number (*see* discussion at **52**), and, a trait it shares only with 1, a factorial equal to itself ($2 = 2!$, *see* **Factorial Family,** pp. 145–146). It is the largest counting number n for which the equation $x^n + y^n = z^n$ has a nontrivial solution, that is, a solution other than $x = y = z = 0$. This is Fermat's Last Theorem (*see also* discussion at **4**).

SIMILAR SPECIES   Multiples of 2 are *even numbers*. All other counting numbers are *odd*. The ancient Greeks showed counting numbers as patterns of dots in sand. They defined evenness as pairing. A number that can be shown as pairs is even, while a number that has a leftover is odd:

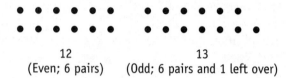

12
(Even; 6 pairs)

13
(Odd; 6 pairs and 1 left over)

Today we also call 0 even and often include negative even numbers. The entire set of positive and negative whole numbers and 0 is labeled the *integers* (from the Latin *integer*, "whole"). Only integers are even and odd. The fraction 2/6 shows two even numbers, but 2/6 has the same value as 1/3, which shows two odd numbers. Many fractions in simplest form are quotients of an even and an odd: 3/4; 2/3.

One could extend the concept of even and odd to finite decimals by looking at last digits, agreeing that 0.1234 is even but 0.12345 is odd; but for numbers shown as infinite decimals, the situation is impossible. Unless the decimal is among the group that repeats the same digit to infinity, such as 2.0000 . . . , 0.3333 . . . , or 1.9999 . . . , it is impossible to know what the last digit of the number is. Furthermore, the number named by 2.0000 . . . is the same as the one named by 1.999 . . . (*see* **Genus Real**), so even knowing the last digit is worthless in determining whether a particular real number is even or odd.

PERSONALITY   Doubling and halving are the easiest operations on their own, and are often used to find other products or quotients. Doubling and adding once is the same as multiplying times 3 (for example, $26 \times 3 = 52 + 26 = 78$); doubling twice is times 4; doubling twice twice is times 8.

Among the most useful sequences to know for mental computation and recognition of important number patterns is that of the first ten powers of 2:

$$2, 4, 8, 16, 32, 64, 128, 256, 512, 1024$$

ASSOCIATIONS The concept of 2 is hidden in words throughout English as well as in other languages. Old English *tweegen* becomes *twi* in words like "twice" or "twain"; "twilight" is "two lights," the times at dawn and dusk when the Sun and Moon are seen together. "Bicycle" and "binocular" use a form of the prefix *bis-* to mean "two." *Twi-* shows up again in "twin" (obvious), "twill" ("two-threaded"), and "twig" ("two branches"). Other 2s are even more hidden, such as in "duel," which is a formal war between two individuals (the Latin *du-bellum* elided into what Lewis Carroll called a portmanteau word). My favorite hidden 2s are "doubt" (hesitating between two alternatives) and "diplomat," the latter from the Roman *diploma*, a sealed set of two bronze tablets given to old soldiers to show that they had certain rights and privileges throughout the Empire. And the two testicles when inflamed have didymitis (*didym-* is another Greek combining form for 2).

There are 2 pints in both the U.S. and the imperial quart and, in the United States, 2 cups to the pint.

It is difficult to locate twoness in geometry. We normally think of geometric figures in terms of either unity or threeness. For example, even a single point separates a line into three regions (two half lines and the point), and similarly a line separates a plane into three regions. Closed figures, such as circles or triangles and other polygons, can be related in one way or another to 1, 3, or some larger integer, but not to

2. There are no identified figures that have just two points, although a line segment consists of two points *and all the points between them*. If two points are different from each other, they have an infinite number of points between them. If the points are the same, then there is just one point, not two.

The number 2 does show up in an unlikely place in geometry. The difference found by subtracting the sum of the number of edges of any three-dimensional figure with flat sides, sharp corners, and no holes (technically a simple polyhedron) from the number of vertices and adding the number of sides (faces) is always 2. For a cube, for example, the number of vertices is 8, the number of edges is 12, so the difference is -4. Adding the number of faces, 6, produces 2. This rather unusual equation

$$v - e + f = 2$$

is named Euler's Theorem, after the Swiss mathematician Leonhard Euler [1707–1783], but was originally discovered by René Descartes [French, 1596–1650].

There is a twoness at the heart of the universe. Quarks are among the most fundamental particles of matter, but they cannot exist alone. Most matter consists of pairs (and trios) of quarks, and there are no ordinary particles formed of four or more joined particles. It is as if God were a member of one of those hunter-gatherer tribes that can count only to 3. Furthermore, each particle can exist in two and exactly two states, known as spin up and spin down. Spin itself is measured in half-integer steps (*see* discussion at **Genus Rational, 1/2**), and there are many particles with a spin of either 1/2 or 1, but only a single particle with spin 2, the still undetected graviton. There is more on this subject at **3**.

The number 2 is a "magic number" for chemistry in two ways. The element with only two electrons (helium) is stable chemically because its outer shell of electrons is filled. The element (or form of an element) with only two nucleons (protons and neutrons collectively) is stable with respect to radioactive degeneration—this is the hydrogen-2 atom, also known as deuterium. Because all the small atoms are fairly stable to begin with, the "magic" stability effect of 2 is less apparent than the chemical and radioactive stability of the elements with higher magic numbers. For chemical stability, these are 10, 18, 36, and 54. For radioactive decay the higher magic numbers are 8, 20, 28, 50, 82, and 126.

Prime Family

Numbers when multiplied with each other are called *factors*. Numbers called *primes* have only themselves and 1 as factors. Numbers that are not prime are called *composites*. The number 1 is not considered to be either prime or composite, so for the first 25 counting numbers the primes are 2, 3, 5, 7, 11, 13, 17, 19, and 23, while the composites are 4, 6, 8, 9, 10, 12, 14, 15, 16, 18, 20, 21, 22, 24, and 25. The importance of the prime numbers in the eyes of mathematicians is commemorated in their having named the following statement, first proved by Euclid, *the fundamental theorem of arithmetic*: any counting number can be factored into primes in exactly one way.

The uniqueness of this result can be used in various ways, which range from simplifying the operation of adding fractions to the proof that the system of natural numbers is either inconsistent or incomplete (Gödel's Theorem, named for its discoverer, the mathematical logician Kurt Gödel [Czech-American, 1906–1978]).

The ancient Greeks, especially the Pythagoreans, looked at the ways that counting numbers could be shown by arranging a set of objects in geometric patterns. Certain characteristics of the numbers could easily be observed from these patterns. Such geometric diagrams in classical times appear to have been drawn in sand or assembled as stones on specially marked counting boards; the printed versions are dot diagrams, which are used throughout this book.

Numbers, properties and connections first emerged from the study of patterns of dots. For example, numbers can be shown either as a set of paired dots or as pairs with one dot left over, which not only creates the classification of even and odd, but also shows that all counting numbers must be either even or odd—the only two arrangements of paired dots possible. Similarly, dot diagrams help make clear if a number is prime or composite. Any composite number of dots can be arranged into one or more rectangles, such as these for 12 (the possible factors are the number of dots on a side):

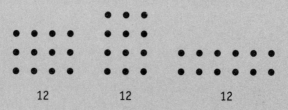

But the nearby prime numbers 11 and 13 cannot be formed into rectangular dot diagrams.

The early Greek mathematicians observed various properties that are related to prime numbers, some of which are collected in books VII and VIII of Euclid's *Elements*. A winner in everyone's list of great theorems is Euclid's proof that there is no largest prime (which is often incorrectly described as the proof that there is an infinity of primes—*see* discussions in **Kingdom Infinity** that explain the difference). Euclid's theorem is what mathematicians call an indirect proof. If you assume that there is a largest prime, then you could make a finite list of all the primes. But there is an even larger number N that is the product of the largest prime with all the primes less than it. Now consider $N + 1$. It cannot be prime, since then it would be in the original list of primes. The alternative is that it must be composite. But it cannot have any of the primes in the original list as a factor, since they will leave a remainder of 1. Therefore, there must be some prime p greater than any in the original list. Euclid's original discussion of the proof is actually more sophisticated, observing that since p divides both N and $N + 1$, it must also divide their difference, which is 1, which is impossible for a counting number.

An important application of prime numbers in today's business world is in the coding (or encryption) of data transmitted by various public means, such as on the Internet. The most common method is based on the difficulty of factoring very large numbers. Even relatively small numbers are not easily factored without a scheme. For example, the identification of all sixteen factors of 210 is of no help when considering the possible factors of the immediate neighbors of 210. Are 209 and 211 prime or can they be factored? If they can be factored, what are the factors? Unless you happen to know the answer from some previous experience with the numbers, the best method for determining factors of relatively small numbers is systematic trial. Divide the number in question by each prime in turn to see if it is a factor. Because of the fundamental theorem of arithmetic, you know that once you have found one prime factor, it will be much easier to unlock the others.

At present (1999) the largest number ever factored is $(2^{15} - 135)^{41} - 1$, a number with 186 digits. It took eighty-eight computers running eighty-eight days to factor it, suggesting that breaking a code based on factoring large numbers will not soon be commonplace, although it may be possible.

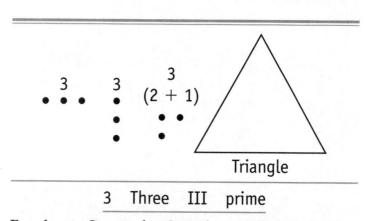

Triangle

3   Three   III   prime

French *trois*; **German** *drei*; **Spanish** *tres*; **Italian** *tre*; **Latin** *tres, tria*; **Greek** *treîs, tria*; **Papuan body counting,** right middle finger; **classical and medieval finger numerals,** left little, ring, and middle fingers folded at joints, index finger pointing, thumb raised

For most of human history an unquestioned fact of life has been that the universe consists of three dimensions, no more and no fewer (but *see* discussions at **4, 5, 10, 11,** and **26** for revisionist views of this idea). Indeed, a 3-D image is much more realistic than a page of this book or your television screen.

The use of 3 as a maximum in various contexts (three strikes and you're out, for example) may go back to the earliest languages and even to the wiring of the human brain. Linguists have noted that in languages that inflect nouns and modifiers, "one," "two," and "three" are the only number words that are inflected. Older languages in particular seem to have had number words up to "three" and no further. The Indo-European base language from which all modern European and Indian languages descend appears to have used the word "three" simply to mean "a lot." Indeed, the phrase "one, two, three" probably meant "one, another one, a lot of ones." So the special place for 3 in the mind is that it is the largest of the old numbers, the ones that humans have always had.

FIELD MARKS　The smallest odd prime number is 3, and 3 is also the larger member of the only pair of consecutive primes. It is also the smallest member of what are known as the *prime twins*, which are prime numbers whose difference is 2, such as 5 and 7 or 11 and 13. There appear to be an infinite number of prime twins, but efforts to settle this question by proof have so far come to naught. In the case of 3, its twin is 5. Furthermore, the numbers 3, 5, 7 form a prime triplet—in fact the only prime triplet, since if there are three consecutive odd numbers, one of them must be divisible by 3.

Other recognizable traits: 3 is a triangular number (*see* **Triangular Family,** pp. 44–47), a member of the smallest Pythagorean triple (*see* **Pythagorean Family,** pp. 67–70), a Mersenne prime (*see* **Mersenne Family,** pp. 85–87), and a Fibonacci and a Lucas number (*see* **Fibonacci and Lucas Families,** pp. 120–122). Indeed, 3 is the only number other than 1 that is both a Fibonacci and a Lucas number.

A hexagram is a hexagon placed in the graph of a conic section (circle, ellipse, hyperbola, or parabola) so that the vertices of the hexagon are all on the graph of the conic—in common mathematical shorthand, a hexagon inscribed in a conic. Pascal's mystic hexagram (*see* discussion at **15**) is mystic because the extensions of the opposite sides of the hexagram always meet in three points that lie on a line (the hexagon inscribed in a circle is the limiting case). Blaise Pascal [French, 1623–1662] and other mathematicians have found many other remarkable geometric configurations involving specific numbers of points and lines that derive from the hexagram.

SIMILAR SPECIES

Many characteristics of 3 also apply to 9. It is easy to remember that 9 is 3×3 as it is that 4 is 2×2. Often we handle a problem that involves 9 or 4 by cutting it in half, so to speak, solving two easy problems that use 2 or 3 instead of a single harder one that uses 4 or 9. One way to divide by 9, for example, is to divide twice by 3. This same approach could be used for larger numbers, but it does not come so naturally.

PERSONALITY

In a numeration system based on 10, a factor of 3 is almost as easy to recognize as a factor of 2, although a completely different method is employed. The last digit of a number is useless, since any digit can appear as the last digit for a multiple of 3 (for example, 12 ends in 2, 15 in 5, 18 in 8, 21 in 1). It is always true, however, that the sum of the digits of a multiple of 3 is also a multiple of 3, and that if the sum of the number's digits is not divisible by 3, then the number cannot be. Therefore, you can look at a large number, such as 123,456,789, and quickly compute the sum

$$1 + 2 + 3 + 4 + 5 + 6 + 7 + 8 + 9$$

as 45, which is easily spotted as multiple of 3: 3×15. In case you don't recognize it, you could add $4 + 5$ from 45 to get 9, which you surely know is a multiple of 3.

Notice that this means that you can scramble the digits of any multiple of 3 and the result will still be a multiple of 3. For example, 987,654,321 and 123,789,456 are both multiples of 3.

ASSOCIATIONS Among the mysteries of Christianity is the Trinity, God in three Persons. This doctrine combines threeness with oneness in a way that remains unlike nearly anything else except the concept of dimension. The three dimensions of length, breadth, and depth are combined into a single space with each dimension both separate and inextricably bound to the other two. The three dimensions of traditional space become two when three points are chosen to determine a plane. Any three points that are not on a line not only determine a particular plane but also form the vertices of a triangle. Knowing the distances between the three points tells everything there is to know, from a geometer's point of view, about a triangle. The angles of a triangle, for example, are determined by the sides, which is the basis of trigonometry (or "three-angle measuring" in Greek-derived Latin). This uniqueness makes the triangle the only rigid figure formed from straight line segments and, therefore, it is vital to engineering. A triangle is rigid because once the vertices are connected, the figure cannot be deformed without bending a side, unlike a rectangle, for example, which can be deformed into a parallelogram. Any polygon with more than three sides can be broken into nonoverlapping triangles (*see* discussion at **14**). Ironically, despite these associations between 3 and angles, it is famously impossible to trisect an angle with straightedge (an unmarked ruler) and compass alone.

One of the words frequently encountered in mathematics is "trivial," which is used to describe instances of a theorem that fit the definition but that are not important. Often the case $n = 0$ or $n = 1$ is a trivial instance of some general rule or type of number. The word is built on the *tri-* prefix that derives from the Latin for "three". Most word mavens say that "trivial" comes from "three ways" (*tri* and *via*, Latin for "road"), meaning the place where three roads meet, an inconsequential settlement; *trivialis* meant "of the street," therefore "commonplace." At least one authority, however, deriving the meaning from the same Latin *trivium* for "three roads," thinks it entered the language from the collective name for the three lesser medieval studies of grammar, rhetoric, and logic, which constituted the courses the university student had to master before embarking on the quadrivium of arithmetic, geometry, astronomy, and music. Hence, "trivial" knowledge is not worth much, as it is only what a beginning student knows.

In knot theory, an important part of topology, knots are classed by the minimum number of times that the rope (which is connected at both ends to form a closed curve) will cross over itself when flattened almost into the plane. Looping is not enough: the knot must be a true knot that can be undone only if the rope is cut, so there are no knots with only one or two crossings (the circle is considered the null knot). So 3 is the minimum number of crossings that produces a knot, and there is only one possible knot with three crossings, the overhand or trefoil knot. This knot comes in mirror images, which are topologically distinct since one cannot be deformed into the other without leaving three-dimensional space for at least four dimensions (not counting time as a dimension in this context). Topologists are perhaps embarrassed by mirror-image knots, but they do not count them as separate (*see also* discussion at **16**).

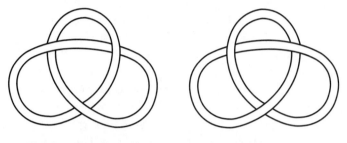

Right - Handed Trefoil   Left - Handed Trefoil

The number 3 seems to be intimately involved with the structure of the universe. In addition to the three dimensions of traditional space, there are according to the standard model of particle physics exactly three families of particles. There is strong evidence from particle accelerator experiments to support this conclusion, although it does not have any theoretical basis. The first family, which includes the electron, a neutrino associated with the electron, and two quarks whimsically named up and down, would seem to provide all the particles a person needs, especially when you add in the antiparticles for each of these. All matter in the universe today is built from particles in the first family.

So it was rather astonishing in 1946 when cosmic "rays" (which are actually streams of particles) were found to contain particles that do not belong to the first family. Although at the time physicists were more surprised by finding particles made from what we now recognize as second-family quarks, in retrospect the oddest discovery was the muon, which is the second-family analog to the electron. Although muons appear easily whenever energies are sufficiently high, the muon seems to have no role or reason for being. It is simply a fat electron, with a mass about 200 times that of the electron. Later, still higher energies in particle accelerators produced what might be called a fat muon, the tauon. Hence, the three families are often called the electron family, the muon family, and the tauon family. Each family also includes its own neutrinos and quarks.

It could be expected that all you would need to find a fourth family would be the addition of still more energy, but experiments seems definitely to rule out this possibility. So, there are three families of matter and no one so far knows why.

The old British measure of a sack was (most of the time) three bushels.

Triangular Family

The followers of Pythagoras developed a classification scheme for numbers based upon their dot diagrams and determined the properties that went with members of particular classes. Prime and composite numbers can be recognized by rectangular patterns of dots for composites and the lack of rectangles for primes. Similarly, the patterns arranged in triangles or other shapes were recognized early. For the most part, the study of numbers that relate to shapes, called *figurate numbers*, does not have extensive practical application, although if you are aware of these patterns, it can be helpful in special situations.

The simplest closed figure with straight sides is the triangle, so studies of figurate numbers begin with triangular numbers. These are formed by adding the following counting number to get the next triangular number, which forms a series of equilateral triangles (all sides show the same number of dots):

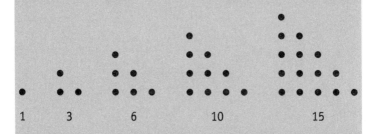

If you count the dots, the numbers shown are 1, 3, 6, 10, and 15. Using a small triangle raised and to the left, like an exponent, we can describe the first five triangular numbers as $1^\triangle = 1$, $2^\triangle = 3$, $3^\triangle = 6$, $4^\triangle = 10$, and $5^\triangle = 15$. It is an interesting challenge to find a formula for $n^\triangle$ if you do not already know it. (If you want to try, then do not read the next paragraphs until you have succeeded or given up.)

The triangular numbers are the sequence formed as 1, 1 + 2, 1 + 2 + 3, 1 + 2 + 3 + 4, and so forth. One way to reason out the formula begins with the idea that if these were shown as right triangles, the height would be the same as the base.

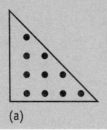

(a)

In this case, height and base each would be 4 units. For a right triangle, the length of one of the sides of the right angle is the same as the height, so the area $A = \frac{1}{2}bh$ of the triangle is half the product of its sides. In this instance, the area (or "dot area") would be half of 4×4, or 8. But the number of dots is 10: the triangular number $1 + 2 + 3 + 4 = 10$, not 8. So this idea does not work. But it does suggest another idea. The area formula for a right triangle is based on the idea that a right triangle is half a rectangle:

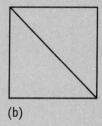

(b)

But if you double the dot triangle to make a dot rectangle, it looks like this:

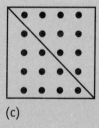

(c)

This is not a 4-by-4 square of dots, but a 4-by-5 rectangle of them. So its dot area should be $A = lw$ or length times width, which is $4 \times 5 = 20$. Half of 20 is 10, which is the correct value for the fourth triangular number $1 + 2 + 3 + 4 = 10$. A little thought will show that this same procedure would work for any triangular number, since each can be shown in a right-angle form that can similarly be doubled. In each case the width of the rectangle will be the same as the number whose triangle is being sought, while the length (= height in these diagrams) is 1 more than that number. Then the dot area of the rectangle is length times width, while the dot area of the triangle is half that amount. So the general formula for triangular numbers is $n^\triangle = \frac{1}{2}n(n + 1)$.

When I earlier computed the sum of the digits in 123,456,789 to see if it was a multiple of 3, I cheated and did not actually add $1 + 2 + 3 + 4 + 5 + 6 + 7 + 8 + 9$. I knew that the triangle of 9 is 45 and I checked to make sure by using the triangular-number formula: $9 \times 10 = 90$; $90 \div 2 = 45$. Although chances to use the triangular numbers do not occur very often in computations, they do occur, and the formula can be useful.

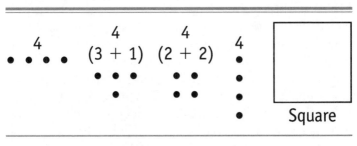

4   Four   IV   2 × 2

French *quatre;* **German** *vier;* **Spanish** *quatro;* **Italian** *quattro;* **Latin** *quattuor;* **Greek** *téssares;* **Papuan body counting,** right middle finger; **classical and medieval finger numerals,** left ring and middle finger folded at joints, index and little fingers pointing, thumb raised

As a child I wondered why the hero of fairy tales searched the four corners of the world; everyone knows that the world is round and has no corners. Of course, the appearance of 4 and the implied angularity in this context echoes the four cardinal directions, north, east, south, and west. The question then occurs, why four directions and not, say, six, or seven? The answer is in the stars and *not* in ourselves. Watchers of the night sky see the entire universe turn about a point. The direction of that point is clearly important. In the hemisphere where our Western astronomical heritage developed, that direction is north. The way that the stars turn is perpendicular to a line from the observer to the magic point in the sky, a point now almost exactly occupied by Polaris, the North Star. Hence, one direction. The stars appear to travel from east to west, just as the Sun and Moon do. This accounts for two more directions, and the underworld must also turn about a point, the point opposite to Polaris, the south pole of the heavens. Hence, four.

FIELD MARKS Every map on a plane can be colored with four colors in such a way that no regions with a common boundary (greater than a single point) will have the same color: this is 4's most dramatic field mark. The same condition applies to maps on a sphere, so it is true of maps on a globe representing Earth as well. This property of 4 was conjectured in 1850 by a schoolboy, Francis Guthrie, who was making a map of the counties of England. When it came to the attention of scholarly mathematicians, every effort to prove it true or false failed, although a number of similar theorems were proved about other shapes, such as the number of colors needed for a map on a doughnut (which, incidentally, is seven). The property became famous as the Four-Color Problem. Finally, in 1976, in the first triumph of the computer over the human mind, mathematicians used a computer program to investigate all possible maps on the plane and sphere and determined that they could indeed be colored with just four different colors. Mathematicians were not thrilled with this proof by what is derisively nicknamed "brute force" in mathematical circles, but they have not yet improved upon the computer's work.

Among number people, the most famous field mark of all is that every number 4 or greater can be written as the sum of no more than four squares. If you include 0^2, the rule becomes: each number 4 or greater can be written as the sum of exactly four squares. This generalization of the idea of a Pythagorean triple (*see* **Pythagorean Family,** pp. 67–70) is far from obvious or easy to prove. Some numbers can be written as the sum of *two* squares; the smallest such is $2 = 1^2 + 1^2$ (or $1 = 1^2 + 0^2$ if you include the squares of 0). Clearly, you can easily annex an infinite list of other examples, beginning with $5 = 1^2 + 2^2$, $8 = 2^2 + 2^2$, and $13 = 2^2 + 3^2$. But the list 2, 5, 8, 13, has gaps (incidentally, note that these sums of two squares are all members of the Fibonacci Family—*see* **Fibonacci and Lucas Families,** pp. 120–122—is this always the case?). Even including 0 will give you the list that starts 1, 2, 4, 5, which passes over 3. So not all natural numbers can be represented as the sum of two squares, even if you include the square of 0.

Similarly, you can make a list of numbers that are the sum of three squares, beginning $3 = 1^2 + 1^2 + 1^2$, $7 = 1^2 + 1^2 + 2^2$, and $9 = 1^2 + 2^2 + 2^2$, but this also has gaps that cannot be filled. But starting with $4 = 1^2 + 1^2 + 1^2 + 1^2$ the list (including the square of 0) begins $4 = 1^2 + 1^2 + 1^2 + 1^2$, $5 = 0^2 + 0^2 + 1^2 + 2^2$, $6 = 0^2 + 1^2 + 1^2 + 2^2$, $7 = 1^2 + 1^2 + 1^2 + 2^2$, $8 = 0^2 + 0^2 + 2^2 + 2^2$, $9 = 0^2 + 1^2 + 2^2 + 2^2$, $10 = 1^2 + 1^2 + 2^2 + 2^2$, and so forth. The mathematician and poet Claude-Gaspar Bachet, Sieur de Méziriac

[French, 1581–1638], verified this kind of decomposition of counting numbers into the sums of four squares through the number 120, and conjectured that the relation was true of all counting numbers. Pierre de Fermat [French, 1601–1665] claimed a proof that was never revealed to his correspondents, so it was not until 1770 that another aristocrat, Comte Joseph-Louis Lagrange [French, 1736–1813], provided the first published proof.

Not only is 4 the smallest number that is a nontrivial perfect square (*see* **Square Family,** pp. 56–59), but it also has another association with perfect squares. Since all counting numbers are either even or odd, they are divisible by 2 or have a remainder of 1 when divided by 2; all perfect squares are either divisible by $2^2 = 4$ or have a remainder of 1 when divided by 4.

The number 4 is the smallest number that can be partitioned by addition (which is called "partitioning" in number theory) to produce more partitions than the number itself. Four of the partitions are symbolized in the dot diagrams for 4 in the heading above. These are $1 + 1 + 1 + 1, 3 + 1, 2 + 2$, and 4 itself; there is also a fifth partition, $2 + 1 + 1$. The lesser numbers 1, 2, and 3 have only as many partitions as themselves—for 3, for example, they are $1 + 1 + 1, 1 + 2$, and 3. Notice that commuted, or reflected, partitions are ignored: $1 + 2$ and $2 + 1$ are counted as a single partition.

All numbers greater than 4 have more partitions than themselves. For example, 5 has 7 partitions, 6 has 11, 7 has 15, and so forth. The number of partitions continues to increase. By the time you reach 200 the number has soared to 3,972,999,029,388. No one has found a general rule for the nth partition, although there exists a rule for finding the $(n + 1)$th from the nth. The great Indian mathematician

Srinivasa Ramanujan [1887–1920] discovered several rules for obtaining partitions for numbers divisible by 5, 7, and 11, although not for all numbers.

Other recognizable traits: The number 4 is a member of the smallest Pythagorean triple (*see* **Pythagorean Family,** pp. 67–70), and the only Lucas number other than 1 that is also a perfect square (*see* **Fibonacci and Lucas Families,** pp. 120–122). Every algebraic equation of degree 4 or less can be solved by algebraic means, but not all those of 5 or greater.

SIMILAR SPECIES   As the first consequential power of 2 (taking $2^1 = 2$ as trivial), 4 is near the start of a family that has a unique place among the counting numbers. Numbers that are powers of 2 figure in many families of numbers and in practical applications (*see* **Powers of 2,** p. 171).

PERSONALITY   After multiples of 2, multiples of 4 are the easiest to recognize. By the time you count by 4s to 20, the pattern of numerals is established and will thereafter repeat:

$$4, \quad 8, 12, 16, \quad 20,$$
$$24, 28, 32, 36, \quad 40,$$
$$44, 48, 52, 56, \quad 60,$$
$$64, 68, 72, 76, \quad 80,$$
$$84, 88, 92, 96, 100$$

When the sixth such cycle begins after 100—

$$104, 108, 112, 116, 120,$$
$$124, 128, 132, 136, 140,$$

and so forth—it is easy to see that the whole pattern from the first five cycles will repeat with the digit 1 in front of it.

As a result, you need look only at the last two digits of any counting number in the 100s to observe whether or not it is a multiple of 4. For example, 722 is not a multiple of 4 (since 22 is not), but 972 is (since 72 is). Furthermore, the cycle continues to repeat for any number of hundreds. No matter how large the counting number is, only the last two digits are needed for you to recognize whether or not is it a multiple of 4. This task is made easier because odd numbers are easily discarded. For speed in recognizing multiples of 4 it is useful to memorize the multiples from 40 through 96. Although these are not part of the basic multiplication facts taught in elementary school, they are very helpful to know.

ASSOCIATIONS   The Four Horsemen of the Apocalypse in the sixth chapter of Revelation are identified simply by the colors of their horses, white, red, black, and pale, although the rider of the pale horse is named as Death. The rider on the white horse is thought to be Christ, and it seems fairly clear that the rider of the red horse is War. The rider of the black horse is associated with food, and thus is often taken to be Famine. Popularly, the four riders are usually called War, Famine, Pestilence, and Death.

The four elements of Aristotle emerge at the end of a long debate among early Greek philosophers as to the nature of the universe. Thales [Ionian, 624–546 B.C.], often considered the first scientist and also the inventor of mathematical proof, proposed that water is the fundamental substance. Anaximenes [Ionian, c. 570–c. 500 B.C.] thought it was air. Xenophanes [Ionian, c. 570–c. 480 B.C.] proposed earth. Pythagoras [Ionian, c. 560–c. 480 B.C.] insisted on number as the foundation of everything, a more reasonable viewpoint from a modern perspective. Heraclitus [Ionian, c. 540–c. 475 B.C.] put his money on fire as the element from which all others were created. It was Empedocles [Hellenic

(Sicilian), *c*. 592–*c*. 432 B.C.] who merged earth, air, fire, and water as the "four elements" we know today, but Aristotle popularized the idea. Since Aristotle was no fan of the Pythagoreans, the idea of number as an element was not a part of the synthesis (but *see* discussion at **5** for a fifth Aristotelian element).

Liquid measure contains a recurring 4, most familiar today from the 4 quarts in a gallon and 4 cups in a quart. In both U.S. and imperial measures, a pint also contains 4 gills—but because the British pint (20 ounces) is larger than the American (16 ounces), the British gill is larger too.

Although until early in the twentieth century everyone accepted that there are exactly three dimensions (*see* discussion at **3**), Albert Einstein [1879–1955] proposed his special relativity theory in 1905, suggesting that time must be considered a fourth dimension. Two years later, Hermann Minkowski [German, 1864–1909], who had been one of Einstein's mathematics teachers, reformulated special relativity in terms of four-dimensional space. In the physical world of relativity theory, time (t) does not behave exactly like the other three dimensions (x, y, and z). Einstein was skeptical at first of four-dimensional space. But he came to see that any point in the universe can be located as (x, y, z, t) and its path through the universe is best understood in terms of these four dimensions (*see* discussions of dimension at **5**, **10**, **11**, and **26**.)

The figure-eight knot is the only knot with a minimum of four crossings.

A problem similar to the Four-Color Problem, first added to mathematical lore by Ferdinand Möbius [German, 1790–1868], was originally suggested to him by his friend

Adolph Weiske. A king's will divides his kingdom into five domains, one for each of for his five sons, with the provisos that each part must border on all the others and that the castle of each son should have a direct road to each of the other castles without in any case crossing the domain of a third son. Try to create such a division, and you will find that the two conditions cannot both be fulfilled. That is not the case for four sons, however. So, to generalize, 4 is the greatest number of bordering domains that each have an interior point which can be connected to an interior point within the other domains without crossing a third domain. This is true of domains in a plane or on a sphere, but not on a torus (a doughnut-shaped surface) or other figures that have "holes."

Physicists have determined that there are only four fundamental forces, although frequently evidence of a fifth force is announced. Originally these four forces were named as strong, weak, electromagnetic, and gravity, but later evidence has changed this slightly. The electromagnetic force is familiar as electric charge and magnetism. The force that holds the nucleus together against the opposing force of electromagnetism was labeled "strong." Gravity is actually the weakest of the four forces, but a different nuclear force, which takes part in certain kinds of radioactive decay, is called "weak." When quarks were discovered, physicists realized that the strong force as they pictured it was not itself fundamental but a side-effect of a different fundamental force that they termed "color" for no really good reason. To complicate matters even more, the electromagnetic and weak forces are, at high enough energies, unified into a single "electroweak force," bringing the number back down to three. Physicists are trying to develop Grand Unified Theories (GUTs) that would combine the electroweak and color forces into a single force, and even a Theory of Everything to bring gravity into the fold. These efforts have not reached fruition, so physicists still deal most of the time with four basic forces. Candidates for a fifth force are generally some form of antigravity. Surprisingly, new developments in cosmology are making the antigravity force seem much more likely than then has been the case until recently.

Where does this leave the number of forces? The most likely number for the immediate future is still four—color, electroweak, gravity, and antigravity.

The famous conjecture known as Fermat's Last Theorem, now proved, has a less well known cousin that is a conjecture of Leonhard Euler. Euler suggested that since it is known that the equation

$$a^n + b^n = c^n$$

can be correctly stated for two well-chosen, distinct counting numbers a and b and $n = 2$, but apparently not for $n > 2$ (Fermat's Last Theorem), and also since the equation

$$a^n + b^n + c^n = d^n$$

can be correctly stated for three well-chosen, distinct counting numbers a, b, and c and $n = 3$ but not for $n > 3$, then perhaps it is the case that for all distinctive nth positive powers, where n is greater than 1, at least n distinct positive nth powers are required to sum to an exact positive nth power. The next in the series would be the sum of four fourth powers needed to equal a single fourth power, followed by five fifth powers required for a single fifth power. But this conjecture is false when $n = 5$, for it was shown in the 1960s that

$$27^5 + 84^5 + 110^5 + 133^5 = 144^5$$

is true, meaning that only four fifth powers are required to sum to a fifth power.

Square Family

Since the early Greeks introduced figurate numbers (*see* **Triangular Family** (pp. 44–47), only one type of them has entered the mainstream of mathematics. Squares—square numbers, that is—are numbers whose dot diagrams are geometric squares. They are formed by adding the odd numbers in sequence: $1, 1 + 3, 1 + 3 + 5, 1 + 3 + 5 + 7$, and so forth. Now each odd number has the algebraic form $2n + 1$ where n can be 0 or a counting number. You could also write $2n + 1$ as $1 + n + n$. Thus, any odd number of dots can be shown as a sort of L-shaped figure that has one dot at the angle of the L and n dots in each arm. Greeks called such figures "gnomons" after their L-shaped sundials. Here are the first few odd numbers shown as gnomons:

1	$1 + 1 + 1 = 3$	$1 + 2 + 2 = 5$	$1 + 3 + 3 = 7$	$1 + 4 + 4 = 9$

It is easy to see that the first gnomon, 1, will fit into the empty space of the L in the second gnomon to make a square with two dots on a side. Thus, $1 + 3 = 2^2$. Similarly, that square fits into the next L to make a square with three dots on a side, or $1 + 3 + 5 = 3^2$. Continuing, $1 + 3 + 5 + 7 = 4^2$ and $1 + 3 + 5 + 7 + 9 = 5^2$. This kind of argument was enough for the Greeks and should be enough for you as well.

(b)

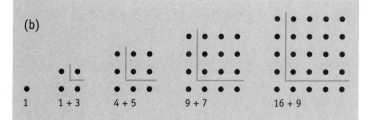

| 1 | 1 + 3 | 4 + 5 | 9 + 7 | 16 + 9 |

The conclusion to be reached is easy to express in modern notation: for any counting number n the formula

$$1 + 3 + 5 + 7 + \ldots + (2n - 1) = n^2$$

is true. Notice that this formula uses a $2n - 1$ formula for odd numbers instead of $2n + 1$.

The square of any counting number is also called a *perfect square* to distinguish numbers such as 1, 2, 4, 9, 16, and so forth (perfect squares all) from numbers such as 3 (the square of the square root of 3) and so forth. Every number is the square of some number, but only the squares of the counting numbers, the square numbers of the Pythagorean dot diagrams, are "perfect." But when dealing with the counting numbers, the modifier "perfect" is usually dropped.

Square numbers have many interesting properties. Here is one involving primes and composites where they appear unexpectedly.

It is not necessary to perform all possible trials with small primes to determine that a large number is prime or composite. The rule for determining when to stop dividing depends on square numbers. The rule, for reasons that will become apparent, is easiest to demonstrate with a perfect square, such as $144 = 12 \times 12$. Notice first that since the factors of 12 are 2 (twice) and 3, the factors of 144 are 2 (four times) and 3 (twice). From these you can work out all the pairs of factors that make 144: 1×144; 2×72; 3×48; 4×36; 6×24; $8 \times$

18; 9 × 16; 12 × 12; 16 × 9; 18 × 8; 24 × 6; 36 × 4; 48 × 3; 72 × 2; 144 × 1. Examination of the list reveals a certain symmetry. After 12 × 12 the factors repeat in reverse order, and it is fairly obvious that this will always happen.

Now consider a number that is not a perfect square, such as 145. Its square root is not a counting number, but it is easy to calculate that the square root must be some number that is only a bit larger than 12, since 12 × 12 = 144. Furthermore, it cannot be as much as 13, since 13 × 13 = 169.

So if you are looking for pairs of factors of 145, you know that when you reach 12 in your search, you can stop. Furthermore, if you want to look only for *prime* factors, the same rule will apply, except that now you know you can stop with 11, since that is the greatest prime number less than 13, and so if you have not found a prime by the time you get to 11 you can stop. The actual prime factors of 145 are then found as follows. It is easy to tell that neither 2 nor 3 is a factor of 145, but 5 is:

$$5 \overline{)145} \atop 29$$

Now you can stop, because 5 × 29 = 145 and both 5 and 29 are prime.

Similarly, for 143 the process would be to test 2 (no), 3 (no), 5 (no), 7 (no), and 11 (no). Then you can stop because the square root of 143 must be a bit less than 12, so it is impossible for there to be a different pair that has a prime factor bigger than 11 since it would have to pair with a prime less than 11, and you have checked all the primes less than 11.

This rule greatly simplifies the search for factors of a number. If there is no prime factor, then, as a consequence of the fundamental theorem of arithmetic, there is no natural-number factor at all. To examine the factors of any number, you need but examine the prime factors less than the square root of the number.

For small numbers this is not an onerous task. For larger numbers with hundreds of digits in their decimal representation, even this simplified method of factoring is of little use. Mathematicians are able to test factors of large numbers by methods that are generally too sophisticated and complex to discuss in this book. The problem of finding factors quickly becomes more difficult with increasing size of the number and, although new methods are frequently discovered, the problem of factoring very large numbers or determining if they are prime—for which we know there is always a solution—is one of the most difficult tasks in mathematics.

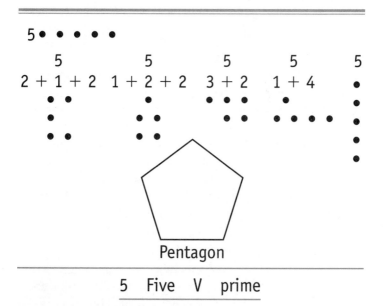

Pentagon

5   Five   V   prime

French *cinq*; **German** *fünf*; **Spanish** *cinco*; **Italian** *cinque*; **Latin** *quinque*; **Greek** *pénte*; **Papuan body counting,** right thumb; **classical and medieval finger numerals,** left middle finger folded at joints, index, ring, and little fingers pointing, thumb raised

Most people can instantly recognize five things without counting them, but not six or more. This was apparently true of the Creator as well—the number 5 is oddly favored by nature. Think of the five-part body plan of the starfish and its relatives, or the five extensions of the body of a quadruped (four limbs and a head). Flowers have either five petals or four. People have five fingers on each hand, and many creatures also have five toes, and possibly for this reason Roman and Mayan numerals are partly based on 5 (the V for 5 of Roman numerals is thought to represent a hand).

[Some psychologists claim that 4 is the maximum number that can be recognized by glancing at a set of objects, so perhaps recognition of five objects without counting is actually observing one more than four objects, not five of them per se.]

FIELD MARKS In 1968 the American mathematician I. A. Barnett published a brief note about a dozen unexpected places that 5 appears in mathematics. Here are five of them:

1. The curves formed by the intersection of a plane with a cone, the conic sections, which include all the graphs of equations in which x and y have exponents of 2 or less (circles, ellipses, parabolas, hyperbolas), can all be specified uniquely by just five points. Shorthand for this property is: "five points determine a conic."

2. There are five and only five figures in three-dimensional space that have all the faces alike and all the angles at the vertices equal. These are the tetrahedron (four faces), cube, or hexahedron (six

faces), octahedron (eight faces), dodecahedron (twelve faces), and icosahedron (twenty faces). This result was known to the ancient Greeks and remarked upon by Plato, so these figures are often called the Platonic solids, but the formal description of a figure with identical faces and angles is that it is *regular*, and closed figures in three-dimensional space that have flat faces are called *polyhedrons* (sometimes *polyhedra*). A short statement of this property might be: "exactly 5 regular polyhedrons exist." (This is the traditional view cited in hundreds of books, but since 1619 it has been known that Euclid's definition admits to other regular polygons; *see* discussion at **9**.)

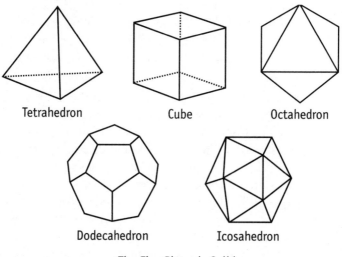

| Tetrahedron | Cube | Octahedron |

| Dodecahedron | Icosahedron |

The Five Platonic Solids

3. The *degree* of an equation in one variable is the same as the greatest exponent for the variable used in constructing the equation. For example, the

degree of an equation such as $3x^2 + 5x - 2 = 0$ is 2 because the highest exponent is 2. "Equations of degree 5 or higher cannot generally be solved by algebraic means."

4. Pierre Fermat was nearly always right, but at least one of his famous ideas was later found to be wrong. He thought that numbers formed by raising 2 to a power that was also a power of 2 and adding 1 would be prime—these numbers are called Fermat primes. The sequence begins:

$$2^0 + 1 = 1, 2^1 + 1 = 3, 2^2 + 1 = 5, 2^4 + 1 = 17, 2^8 + 1 = 257, 2^{16} + 1 = 65,537$$

But the next number, $2^{32} + 1 = 4,294,967,298 + 1 = 4,294,967,299$, is not a prime. So far no other Fermat primes have been found, so "there are only five known Fermat primes" (*see also* discussion at **17**).

5. The foundation of number theory is the study of the integral solutions to equations in which the coefficients are also integers. A relatively simple class of such equations comprises equations in two variables for which each term with a variable is of the third degree and the sum of all these terms is equal to 1. These are such equations as $2x^3 + 3y^3 = 1$ or $x^3 + 2x^2y - 3xy^2 + 4y^3 = 1$. Integral solutions to such equations may or may not exist. A point (x, y) for which both coordinates are integers is called a *lattice point* of the graph. Depending on the graph, such points can be rare or common. Thus, the following number association with 5 is quite remarkable: "the curve $ax^3 + bx^2y - cxy^2 + dy^3 =$

1, where *a*, *b*, *c*, and *d* are integers, passes through, at most, five lattice points."

As a prime, 5 figures in several relationships. It is a Sophie Germain prime (*see* discussion at **2**) and a prime twin with 3 and with 7, so it is the middle member of the only prime triplet. Finally, 5 is the smallest member of a class of prime numbers called Wilson primes that are defined as primes *p* for which 1 more than the product of the counting numbers up to but not including *p* is p^2. For 5, this means that $1 + 1 \times 2 \times 3 \times 4 = 25$, and so 5 qualifies.

Other recognizable traits: 5 is the largest member of the smallest Pythagorean triple as well as of the second Pythagorean triple (*see* **Pythagorean Family,** pp. 67–70), a Fibonacci number (*see* **Fibonacci and Lucas Families,** pp. 120–122), a square pyramidal number (*see* **Square Pyramids,** p. 127), a Catalan number (*see* discussion at **14**), and a Bell number (*see* discussion at **52**).

SIMILAR SPECIES The number 5 and its multiples with powers of 10 are hidden in nearly every measurement. Because 5 is half of 10, it and the multiples of 5 and the powers of 10 are the midpoints used in almost all scientific measurement. If a measurement is reported as 6000 meters, it can have different meanings depending on how the measurement was made. The presence of the three 0s at the end suggests that the measurement was not to the nearest meter, so the true measurement may be half a meter (500 centimeters) less or more. If measured to the nearest 10 meters, the true measure is between 5995 and 6005 meters; to the nearest 100, between 5950 and 6050. Wherever there is a measurement in whole units, there is a hidden 5.

PERSONALITY   The products of any natural number and 5 are among the most recognizable of any numbers in our numeration system: the endings are either 0 or 5.

Since 5 and 2 are the only factors of 10, the numbers 2 and 5 are associated with each other in many computational methods, as are their multiples, especially the multiples of 2 and 5 by 10 and its powers—20, 50, 200, 500, and so forth. Multiplying by 5 in the decimal system is the same as multiplying by 10 and dividing by 2. For example, to multiply 47×5, think $470 \div 2$.

The inverse procedure also is effective for division by 5—that is, instead of dividing by 5, multiply by 2 and move the decimal point one place to the left. Thus, to divide 487 by 5 think twice 487 is 974, so $487 \div 5 = 97.4$. The more digits there are, the more useful the shortcut. When the number of digits in the amount to be divided is too great for a cheap calculator, mental computation is often required. Consider dividing a number such as 398,857,204 by 5. Doubling is easy, either by scribbling down the number twice on a scrap of paper and adding or by starting from the right with 4 and working back as you double mentally and write the digits, also from right to left (thinking 8, 0, 4, 14, $10 + 1 = 11$, $16 + 1 = 17$, $16 + 1 = 17$, $18 + 1 = 19$, $6 + 1 = 7$, while writing 797714408). Then add a decimal point one place from the right and insert appropriate commas to obtain the answer of 79,771,440.8.

This method is even more useful when the divisor is a number such as 50 or 5000. Then the only difference in procedure is that the decimal point in the answer is placed *farther* to the left by the same number of places as the number of zeros in the divisor. For example, $398,857,204 \div 500 = 797,714.408$; because of the two zeros in 500 the answer has its decimal point two places farther to the left than does the answer for $398,857,204 \div 5 = 79,771,440.8$.

ASSOCIATIONS   Clocks combine an Egyptian hour system based on 12 with a Mesopotamian minute and second system based on 60. In the relationship between 12 and 60, 5 is the third partner, since $5 \times 12 = 60$. As a result, the passage of 5 minutes is reported by moving from one numeral to the next on a clock face, making 5 minutes and its multiples convenient measures of time.

In Aristotle's view, the four elements of earth, air, fire, and water (*see* discussion at **4**) were not applicable to the heavens, which seemed to him to have different physical laws from the natural world about us. Thus they must be made from a different element, a fifth element or *quinta essentia*. From this phrase we derive the word "quintessence" to mean something especially pure, and by extension the word is often used to suggest the best or highest.

The success of the Einstein–Minkowski explanation of space and time as a four-dimensional entity inspired Polish physicist Theodor Kaluza [1885–1945] to develop in 1921 a theory in which relativity theory and gravity were combined in a space of five dimensions. Kaluza's five-dimensional universe made sense mathematically but not in terms of the universe we see about us; we experience time and three dimensions of space, but not a fifth dimension. Five years later another physicist, Oskar Klein, worked out a way for the fifth dimension to exist but not be experienced: it is curled up into a tiny cylinder that is too small to be observed—much smaller than the nucleus of an atom. The mathematical theory still makes sense, but few thought at the time that this theory describes the real universe (but *see* discussion at **10**, **11**, and **26**).

In geometry, 5 is most commonly associated with two polygons: the pentagon with five sides and five angles; and the familiar five-pointed star, which is known formally as a pentacle or pentagram (mathematicians say *pentagram* but dictionaries prefer *pentacle*). The pentagram has ten sides and ten angles (five at the points and five between the points), although some authorities insist that it must be constructed to include a second star within. It can be described as the figure formed by connecting the vertices of a regular pentagon and then eliminating both the original pentagon and the smaller one that appears in the middle of the star. The internal vertices of the pentagram can be used to produce the second star by the same procedure. This figure, with or without the second star, is sometimes called the "pentagram of Pythagoras" (for a discussion of the importance of this figure in the history of mathematics, *see* **Irrational Family,** pp. 274–275).

The Pythagoreans' figurate numbers included pentagonal numbers given by the sequence $N = 1 + 4 + 7 + \ldots +$

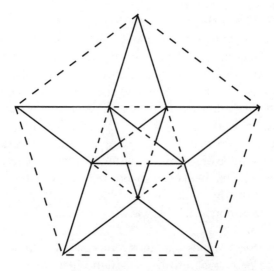

Pentagram of Pythagoras

$(3n − 2) = 1/2 \ n(3n − 1)$. Thus, 5 itself is pentagonal; and the number "5 pentagonal" (or fifth pentagonal number) is 35.

The Four-Color Problem has an associated five-color theorem discovered by Percy John Heawood [English, 1861–1955], who in 1890 proved that five colors suffice to color any map on a plane or a sphere. Unlike the somewhat controversial 1976 computer proof of the four-color theorem, the five-color theorem stems directly from Euler's Theorem (*see* discussion at 2) by ordinary mathematical reasoning.

There are only two knots that have a minimum of five crossings, and there are five knots that have a minimum of six crossings.

There is evidence that in one species of wasp, *Genus eumenus*, the female can count to 5 (and also to 10). She must provide paralyzed caterpillars for a developing larva. Because a female larva requires more food than does the smaller male, the adult female always places five caterpillars in the place where the male egg is laid and ten in the vicinity of a female egg. Some other species of related wasps provide even higher numbers of caterpillars and manage to keep the numbers straight—twelve or even twenty caterpillars are the norm for some solitary wasp species.

Pythagorean Family

Sometimes it seems as if humans are the only creatures on Earth who love straight lines and especially square corners. An archaeologist who uncovers a rock shaped like a box can be almost sure that it is a part of a structure made by humans.

If an astronomer or space probe observed square corners and straight lines on another planet, surely intelligent life—that is, something that thinks like a human—would be suspected. Very early humans seem not to have fallen for the square corner, since early dwellings are mostly circular. But the beginning of civilization can be seen as the time that the right angle appears as the basis for architecture. If you see a photograph of a hunter-gatherer village that has round dwellings, you unconsciously assume that its society is less intricate than an otherwise similar village with square-corner architecture.

One reason we love square corners is that they come out even—four square corners side to side cover a space with no overlapping—although for some purposes angles of 120° work even better, as honeybees seem to have discovered. (Only three of the honeybee angles are needed to cover a space.) If you used parallelograms instead of rectangles to lay out a building or a field, for example, you would need to make two different angles, such as 60° and 120°, or 45° and 135°, but for 90°, only one angle is needed. Furthermore, a closed figure with four right angles will always have its opposite sides parallel and, although this is less obvious, equal in length. Much can be said for the honeybee alternative of the 120° angle and its associated hexagons, but right angles and rectangles—the word *rectangle* means "right angle"—are the human alternative.

Consider the ancient Egyptians. Egyptian mathematics is famous for the "rope stretchers," teams of surveyors who worked to restore property lines after the annual Nile flood. As reported by Greek travelers, they relied on a simple device to lay out square corners. They used a loop of rope knotted with three knots carefully spaced at measured lengths. Each of a team of three stretchers grasped one of these knots and pulled, so the loop became taut. The result was a large triangle in which the largest angle was always a right angle. The square bases of the pyramids with nearly perfect right angles are physical evidence that the rope-stretching method is an effective way to form right angles. Indeed, my wife and I still use this method to produce right angles for fences or in laying out a garden.

For this method to work, the knots have to be correctly spaced along the loop. In the Egyptian spacing, the distances between knots were 3 units, 4 units, and 5 units. In that case, the right angle is between the 3 and 4 sides of the triangle. It does not matter what units are used for measuring the rope, so triangles of any size can be constructed. Using double units or triple units would produce 6–8–10 or 9–12–15 triangles, but these are essentially the same as the 3–4–5 version.

Are there other right triangles that also have three sides that are counting numbers (and do not simply reduce to 3–4–5)? The answer to the question was apparently a preoccupation of another early civilization, since the Old Babylonians of about 1900 to 1600 B.C. left behind clay tablets with lists of the sides of such triangles. The purpose is not clear, but the lists appear to be exhaustive and include such less obvious combinations as the 119–120–169 triangle, in which the right angle is between the 119 and the 120 sides.

It is not certain which early civilization deserves the credit for recognizing that such triangles share a trait that is true for all right triangles: the sum of squares of the lengths of the two short sides is always equal to the square of the side opposite the right angle. The oldest source is Chinese, but the diagram used in that instance applies only to right triangles where the short sides are equal. Certainly the Egyptians knew special cases of this famous relationship, and the Old Babylonians appear to have had a very general view of it. Nevertheless, because it entered Western civilization through Greek sources, we know the relationship today as the Pythagorean theorem. It is possible that Pythagoras did contribute to the development of the theorem by providing a Greek-style proof of it, since proof was not a feature of Babylonian mathematics and only hinted at in Egyptian. Ironically, the followers of Pythagoras used this theorem in a famous proof that refutes the basic Pythagorean idea that everything in the universe is constructed from counting numbers (*see* **Irrational Family,** pp. 274–275).

In any case, because the theorem about the sides of a right triangle, is known to one and all as the Pythagorean theorem, any three counting numbers a, b, and c for which the equation $a^2 + b^2 = c^2$ is true is called a Pythagorean triple. The smallest such triple is 3, 4, 5, the sides of a 3–4–5 right triangle, and also numbers that satisfy the relationship $3^2 + 4^2 = 5^2$. The next smallest is 5, 12, 13.

The Old Babylonians apparently knew that a Pythagorean triple always results from choosing two different counting numbers m and n and using them to define a Pythagorean triple a, b, c by the rule that $a = m^2 - n^2$, $b = 2mn$, and $c = p^2 + q^2$. It is easy to use algebra to show that this will always produce $a^2 + b^2 = c^2$, although it is not clear how the Babylonians worked out this result. When $m = 2$ and $n = 1$, the familiar 3, 4, 5 triple results. Similarly, $m = 3$ and $n = 1$ gives 8, 6, 10, while $m = 3$ and $n = 2$ produces 5, 12, 13.

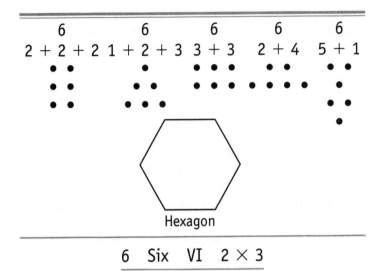

| 6 | 6 | 6 | 6 | 6 |
| 2 + 2 + 2 | 1 + 2 + 3 | 3 + 3 | 2 + 4 | 5 + 1 |

Hexagon

6   Six   VI   2 × 3

French *six*; **German** *sechs*; **Spanish** *seis*; **Latin** *sex*; **Greek** *héx*; **Papuan body counting,** right wrist; **classical and medieval finger numerals,** left ring finger folded at joints, index, middle, and little fingers pointing, thumb raised

Because its multiples have many small divisors, 6 is some-times used in developing systems for writing numbers. Consider 12, with divisors 2, 3, 4, and 6; 24, with 2, 3, 6, 8, and 12; or 30, with 2, 3, 5, 6, 10, and 15. As you can recognize, each new multiple will include more small divisors. The number 60 is particularly rich, with 2, 3, 4, 5, 6, 10, 12, 15, 20, and 30. The first known system of writing numbers, which developed in Mesopotamia, was partly based on 60, and we see its remnants in 60 minutes and 60 seconds. Computer programmers often used a "hexadecimal" system based on a combination of 6 and 10.

FIELD MARKS The number 6 is the least, or first, *perfect number*. A perfect number is one for which the sum of its proper divisors is the number itself. (Proper divisors include 1 but do not include the number itself.) The number 6 qualifies because its proper divisors are 1, 2, and 3, and $1 + 2 + 3 = 6$. Note also that $1 \times 2 \times 3 = 6$. This last statement means that 6 is a factorial number ($3! = 6$; *see* **Factorial Family,** pp. 145–146), which somehow makes 6 seem even more perfect.

Most numbers are less than perfect, and are usually called *deficient* (or sometimes *defective*) because the sum of the divisors, not including itself, is less than the number. Clearly 1 is omitted from discussion by the definitions of perfect and deficient numbers, so 2 is the first deficient number (all primes are deficient) and 4 is the first deficient nonprime. (*See also* discussions of *abundant numbers* at **12** and *multiply perfect numbers* at **120.**)

Somehow the involvement of 6 in the longest perfect Golomb ruler seems related to its status as the first perfect number, although there is no real connection. A Golomb ruler (invented by the American mathematician Solomon W. Golomb in the 1970s) can measure any counting-number length as the distance between two marks without duplicating any distance. A perfect Golomb ruler can measure *all* the counting-number distances up to the length of the ruler without duplication. For instance, a ruler with marks at 0, 1, and 3 can measure not only 1 unit between 0 and 1, and 3 units between 0 and 3, but also 2 units between 1 and 3. Thus, the ruler with marks at 0, 1, and 3 is a perfect Golomb ruler with three marks. It can be proved that there is only one perfect Golomb ruler with four marks, and its length is 6 units. The marks are at 0, 1, 4, and 6. It is an easy exercise to determine that the distances between every pair of marks are 1, 2, 3, 4, 5, and 6. Try it.

The product of any three consecutive counting numbers is divisible by 6. This is true because in any triplet of consecutive counting numbers, one of the numbers must be even and another must be a multiple of 3.

The smallest cyclic number in the Hindu-Arabic system has six digits: 142,857. A *cyclic number* for a given number of digits is one that when multiplied by any counting number less than or equal to the number of digits will have the same digits as the original number but, of course, arranged in a different order. The digits cycle, so starting with one digit the others follow in the same order if you put the last before the first. Note that $2 \times 142,857 = 285,714$; $3 \times 142,857 = 428,571$; $4 \times 142,857 = 571,428$; $5 \times 142,857 = 714,285$; and $6 \times 142,857 = 857,142$. The six digits here can be recognized as the digits that repeat in the decimal expansion of $1/7$ (*see* discussion at **Genus Rational, 1/7**).

The minimum number of colors needed to color a map (*see* discussion at **4**) on a one-sided surface such as a Möbius strip or a Klein bottle is six. This seems odd, since a one-sided surface is like the outside of a sphere in many ways, and the number of colors needed for a sphere is four, the same as for a plane. And the Klein bottle, aside from passing through its own inside, is like a torus, which requires seven colors. Nevertheless, the best authorities all state that for Möbius strips and Klein bottles, it takes at least six colors to make a map.

SIMILAR SPECIES   After 6, the next perfect number is 28, whose divisors are 1, 2, 4, 7, and 14 ($1 + 2 + 4 + 7 + 14 = 28$). After that comes the perfect number 496.

The largest number of divisors possible for a number of four or fewer digits is 62 if you do not count 1, and the two different numbers that have this many divisors—7560 and 9240—are both multiples of 6.

PERSONALITY   Although 6 is not very tractable by itself, it can easily be broken into 2 and 3. Thus, to multiply a number mentally by 6, one good way to do it is to multiply first by 3 and then double the answer. For 37×6, think 37 times 3 is $37 + 37 = 74 + 37 = 111$; 111 doubled is 222. Another approach would be to break 6 into 1 more than its friendlier neighbor 5, so for 394×6, think $394 \times (5 + 1) = (394 \times 5) + 394$; 394×5 is half of $394 \times 10 = 3940$; $3940 \div 2 = 1970$. Then, 394×6 is $1970 + 394 = 2364$. (This is much easier to do than it is to describe.)

Division by 6 can also be accomplished by dividing successively by 2 and by 3. It helps to recognize before you start whether or not the number being divided is a multiple of 6. The easiest way is to use the tests for multiples of 2 and 3 together; for example, 396 is even (so a multiple of 2) and its digits sum to 18, a multiple of 3, making 396 also a multiple of 3, so 396 is also a multiple of 6. Thus, division of 396 by 6 will have no remainder. On the other hand, for $9874 \div 6$, think $9 + 8 + 7 + 4$ is 28, so 9874 is not a multiple of 3 and therefore there will be a remainder. Divide by 2 ($9874 \div 2 = 4937$); divide 4937 by 3 ($4937 \div 3 = 1645 \text{ r } 2$).

ASSOCIATIONS   St. Augustine is among the early Judeo-Christian writers to suggest that God chose six days in which to create the universe because it is a perfect number, and, as Augustine noted, "We must not despise the science of numbers."

The Pythagoreans worked out a series of hexagonal numbers (numbers whose dot diagrams can be arranged as hexagons) as part of their study of figurate numbers. These are found according to the following sequence: $N = 1 + 5 + 9 + \ldots + (4n - 3) = 2n^2 - n$.

The radius of a circle is related to the number 6 in a way that would be surprising if it were not used so often without contemplation. If you begin at a point on the circle and with a compass mark off arcs equal to the radius, after six marks you are exactly back at the starting point. This remarkable relationship between the radius and the circle can be used to construct a regular hexagon or a six-pointed Star of David in a circle.

Why does this occur? It happens largely as a result of 6 being two 3s. A diameter is the same as a straight angle (180°) with its vertex at the center of the circle. The sum of the three angles of any triangle is also 180°. Thus, if all the angles of a triangle are equal, then each measures 60° (180 ÷ 3 = 60). You can always assemble three equilateral triangles on each side of the diameter because the sum of three 60° angles is a straight angle or straight line:

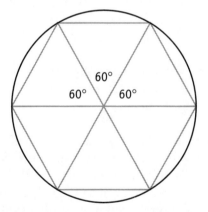

The result is a regular hexagon. Each of the six vertices of the regular hexagon is 1 unit away from the center, so a circle with a radius of 1 unit that has that point as its center will pass through all the vertices of the hexagon. The main reason 6 is involved is that each of the two straight angles has been broken into three triangles, and $3 \times 2 = 6$.

The packing problem for a plane—how many circles of a given size can be arranged around a circle of the same size—is a related challenge. Once again, the answer is 6.

Honeybees seem to know this theorem (*see also* **Pythagorean Family,** pp. 67–70). At least one scientific study found that honeybees create combs with hexagons because they are trying to cram as many cells as possible into a given space. They begin with circles but end up with hexagons as, with the pressure of the stored honey, the circular wax cells merge with one another and share walls. Scientists have long known that this system provides the maximum amount of honey for the minimum amount of wax. While such thrift was once ascribed to the benevolent wisdom of God and at another time to the cunning of the insect, the evidence seems to suggest that any time essentially round creatures pack themselves together in storing something, they will by geometry arrive at the honeycomb structure.

The figure corresponding to a regular hexagon in three space is a *cube*, which has six sides that are all squares. This is the only six-sided regular polyhedron.

There are six raised dots in the Braille system of notation used by many blind people to read and write. It is named for its inventor, Louis Braille [French, 1809–1852], blind himself, who devised it at the age of fifteen. Each sign is shown as a pattern of 6 raised dots that can be read with the fingertips. There are sixty-four possible places to have a dot or not in an arrangement of six. The number 64 is 2^6, and arises in this situation because for each of the six positions (arranged in two columns of three dots each) there are two choices, whether to have a dot or not. For obvious reasons Braille omitted the choice of no dots at all, leaving sixty-three symbols. For ease in reading, Braille used the top pair and the middle pair for the first letters of the alphabet, then added the bottom dots to form the less common later letters—W, which is not at all common in French, was originally omitted. Some of the later combinations were used for diphthongs, indicators, or frequently used words. Numerals, at least in the version used today, are indicated by preceding one of the first ten alphabet letters with a special sign:

so the numerals for 0 through 9 are as follows:

| 0 | 1 | 2 | 3 | 4 | 5 | 6 | 7 | 8 | 9 |

Note that the numeral 0 corresponds with the letter J because Braille started with A equal to 1.

There are just five knots with a minimum of six crossings.

A *siesta* takes place at the sixth hour, which, when the daily round was divided into twelve parts of day and twelve parts of night, no matter the season, would be the middle of the day.

Pascal's Triangle

Notice that if you know the prime factors of a number, it is easy to reconstruct the other factors. For example, knowing that $2 \times 3 \times 17 = 102$ also tells you that the only possible combinations (ignoring the trivial factor of 1) that produce 102 in addition to $2 \times 3 \times 17 = 102$ are $(2 \times 3) \times 17 = 6 \times 17 = 102$; $2 \times (3 \times 17) = 2 \times 51 = 102$; and $(2 \times 17) \times 3 = 34 \times 3 = 102$. In the nineteenth century, various mathematicians experimented with generalizing this simple idea. For example, the number of possible combinations for three factors consists of the three combinations of two factors as well as the single combination of all three; so every number that has three prime factors can be factored into four different combinations. But suppose that there are four prime factors, say 2, 3, 5, and 7. For this situation you need to consider all four prime factors, every combination of three of the primes, and finally the results of pairing factors.

You would begin with the single combination of all four: $2 \times 3 \times 5 \times 7 = 210$.

To determine how many combinations there are of factors taken three at a time, recognize that each such combination can be found by isolating exactly one of the other factors, so there are four combinations, one for each factor:

$2 \times (3 \times 5 \times 7) = 210$ $3 \times (2 \times 5 \times 7) = 210$
$5 \times (2 \times 3 \times 7) = 210$ $7 \times (2 \times 3 \times 5) = 210$

or

$2 \times 105 = 210$ $3 \times 70 = 210$
$5 \times 42 = 210$ $7 \times 30 = 210$

It is a little more difficult to be certain that you have all the combinations that are pairs, but the concept is somewhat the same. Instead of isolating each of the individual factors, you pair each factor, one at a time, with each of the other three factors, leaving the remaining two factors as a second pair. For example, you pair 2 with 3, leaving 5 and 7 to obtain

$$(2 \times 3) \times (5 \times 7) = 6 \times 35 = 210$$

Then pair 2 with 5 to obtain

$$(2 \times 5) \times (3 \times 7) = 10 \times 21 = 210$$

and 2 with 7 to obtain

$$(2 \times 7) \times (3 \times 5) = 14 \times 15 = 210$$

Repeating the process beginning with 3, 5, or 7, however, produces no new combinations, because you have already considered the pairs of each of these in previous combinations.

There is just one combination with all four prime factors, four with combinations of three factors, six ways to use pairs of factors, and the original four prime factors considered as separate numbers, a pattern of totals that can be listed as 1, 4, 6, 4. Thus there are fifteen different factors of 210 in all—sixteen if you count 1, which is a factor of every number. These are 1, 2, 3, 5, 6, 7, 10, 14, 15, 21, 30, 35, 42, 70, 105, and 210.

This kind of thinking led not only to a general consideration of the combinations of various kinds but to a whole important branch of mathematics, called group theory, that goes far beyond simple arithmetic.

Those who have some experience in algebra may recognize the pattern of numbers of combinations explored above (this time with the 1 that is the factor of every number included):

1 4 6 4 1

That is 1 (the number itself as a factor of four primes); 4 (three factors); 6 (two factors); 4 (single prime); 1 (the number 1 as a factor). This pattern is the same as the coefficients of the fourth power of a binomial:

$$(a + 1)^4 = a^4 + 4a^3 + 6a^2 + 4a + 1$$

An ancient way of generating such patterns is known as Pascal's Triangle, since it was investigated by Blaise Pascal in 1654; but it is known as "The Old Method" in a Chinese book of 1303 and as early as 1100 was probably known also to Omar Khayyám [Persian, 1048–1131], who is thought to have learned it from even earlier Indian or Chinese writers. The numbers in each row of Pascal's Triangle are generated by adding pairwise the numbers in the row above. Imagine a universe of 0s in which, suddenly, there appears a 1, and a universal law that says that two adjacent numbers when added will produce the number below:

```
            1
          1   1
        1   2   1
      1   3   3   1
    1   4   6   4   1
  1   5  10  10   5   1
1   6  15  20  15   6   1
```

Pascal's Triangle has many interesting applications as well as various patterns buried within it. Two of the more obvious appear by looking down the diagonals:

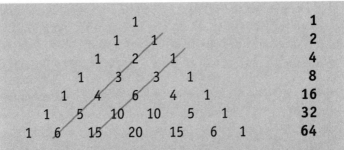

1							**1**
1	1						**2**
1	2	1					**4**
1	3	3	1				**8**
1	4	6	4	1			**16**
1	5	10	10	5	1		**32**
1	6	15	20	15	6	1	**64**

While the first diagonal is simply the counting numbers in order, the second is the triangular number sequence. Furthermore, adding the members of the rows produces the sequence of powers of 2. The sums of another set of diagonals are the Fibonacci numbers (see **Fibonacci and Lucas Families,** pp. 120–122).

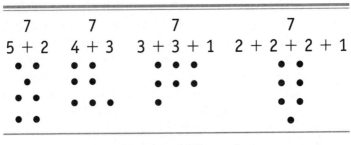

7   Seven   VII   prime

French *sept*; **German** *sieben*; **Spanish** *siete*; **Italian** *sette*; **Latin** *septem*; **Greek** *heptá*; **Papuan body counting,** right elbow; **classical and medieval finger numerals,** left little finger folded at knuckle, thumb raised

It has been suggested that 7 came to be associated with good luck in the Western tradition because of the seven ancient "planets"—the Sun, Moon, Mercury, Venus, Mars, Jupiter, and Saturn—which are also used to name the seven days of the week. Earth was not on the list in ancient days: it was not recognized as a starry wanderer, just a place from which to view the stars. The division of time into weeks of seven days, however, appears to represent the number of days in each phase of the Moon, an association that probably preceded naming the days for the planets. The archaic English name for a week is "sennight," or "seven nights," from the Old English *seofon nihta*, reflecting a Germanic tradition of counting the nights in a passage of time instead of days (*see* discussion at **8**). There are also some important constellations with seven stars, including the star pattern we commonly call the Big Dipper (or in England, the Plough), and more formally Ursa Major.

Of course, a more likely reason that 7 is associated with luck is that the odds of a 7 in dice—which were used in Egypt as early as 2000 B.C., or even earlier—are higher than those of other numbers because there are more ways that a pair from 1, 2, 3, 4, 5, and 6 can add to 7 than to any other sum. The sums $1 + 6$, $2 + 5$, and $3 + 4$ can each occur in two ways in a toss of a pair of dice, making six combinations in all. Compare with the five ways to get a sum of 6 ($1 + 5$ and $2 + 4$ each in two ways but $3 + 3$ in only one way), or the three ways for 8 ($5 + 3$ in two ways and $4 + 4$ in one way).

FIELD MARKS   Division by 7 of any number *not* a multiple of 7 produces, as the various steps of the division algorithm are continued to produce the decimal fraction quotient, all the possible nonzero remainders less than 7. For example,

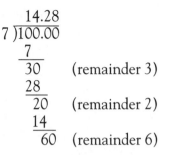

$$\begin{array}{r} 14.28 \\ 7\,\overline{)100.00} \\ \underline{7} \\ 30 \qquad \text{(remainder 3)} \\ \underline{28} \\ 20 \qquad \text{(remainder 2)} \\ \underline{14} \\ 60 \qquad \text{(remainder 6)} \end{array}$$

and so forth. This trait of 1/7 produces a decimal fraction (0.142857142857 . . .) with a repeating period six digits long. Note that 1, 4, 2, 8, 5, and 7 are not the remainders, which must each be less than 7. The repeating period is the cyclic number 142,857 (*see* discussion at **6**), which emerges in a different order for 2/7, 3/7, 4/7, 5/7, and 6/7. For example, 6/7 as a decimal begins 0.857142 and then repeats the period 857142.

Other recognizable traits: 7 is a Mersenne prime (*see* **Mersenne Family,** pp. 85–87), a Lucas number (*see* **Fibonacci and Lucas Families,** 120–122), and a prime twin with 5 and prime triplet with 3 and 5, making it the greatest member of the only prime triplet. It is also part of the third smallest Pythagorean triple 7, 24, 25 (whose members are usually ordered beginning with the smallest leg).

SIMILAR SPECIES Other numbers less than 100 that have all possible remainders during the steps of division, and therefore also generate cyclic numbers, are 17, 19, 23, 29, 47, 59, 61, and 97. There are probably an infinite number of primes with this property, but this has never been proved.

PERSONALITY The number 7 is difficult to use in computation. It is prime, moderately large for a digit, not close to

10, and not a factor of 10 or 100. There is a divisibility rule for 7, just as there is for the other digits, but it is so complex that it is always easier to divide and see if there is a remainder than it is to apply the rule.

Probably because of its computational intractability, 7 and its powers figure in the earliest known example of recreational mathematics. This is Problem 70 of the Egyptian Rhind Papyrus, which dates from 1650 B.C. but is thought to contain material of a much earlier date. Roughly, the problem states: "In each of seven houses live seven cats. Each cat eats seven mice. Each one of the mice would have eaten seven ears of grain had it not been for the cat. One ear, if planted, would have yielded seven bushels of grain. How much grain was saved by the cats?" (The problem is presented as a list used in the solution, so you have to work backward from the solution to guess at the problem.) The solution, from the Egyptian scribe Ahmes, is correct: 16,807 bushels of grain. This same idea is behind a problem created nearly 3000 years later by Fibonacci, and also is implicit in the trick question of a children's rhyme:

> As I was going to St. Ives,
> I met a man with seven wives.
> Every wife had seven sacks,
> Every sack had seven cats,
> Every cat had seven kits:
> Kits, cats, sacks, wives—
> How many were going to St. Ives?

I trust you know the correct answer, but perhaps were fooled once when you were young.

ASSOCIATIONS   The Bible is filled with 7s. The vision of St. John the Divine, described in Revelation, is a rich example: there are the seven churches that are in Asia matched with the seven Spirits (angels) before God's throne, later identified with seven golden candlesticks and seven stars and provided with seven trumpets, seven vials of the wrath of God, and seven plagues; here too are the seven seals on the scroll that can be opened by the Lamb with seven horns and seven eyes.

Scholars in the Middle Ages determined that there are Seven Deadly Sins (usually given as pride, lust, envy, anger, covetousness, gluttony, and sloth). There is a corresponding, but lesser known, group of Four Cardinal Virtues (prudence, justice, fortitude, temperance) and Three Theological Virtues (faith, hope, love)—which of course add up to 7.

One of the most famous problems in mathematics since the latter part of the nineteenth century dates from a 1735 proof of Leonhard Euler that there is no way to cross all the seven bridges that span the River Pregel in Königsberg by a path that crosses each bridge exactly once. This problem and its proof of impossibility are considered the beginning of a significant branch of mathematics known somewhat confusingly as graph theory ("confusingly" because it has nothing to do with any kind of graph) and also, in a happier formulation, as network theory.

There are just eight knots that have a minimum of seven crossings.

Mersenne Family

The new sociologists of science tell us that before the rise of scientific societies and scientific journals, scientific truth did not exist. While few scientists believe that this analysis is correct, nevertheless it is true that early scientific or mathematical ideas could easily perish with the death of the discoverer. It is difficult to picture a scientist or mathematician of today finding something clearly important and failing to publish it, as Pierre Fermat famously did not publish his Last Theorem. Fermat was a country lawyer in provincial Toulouse who in fact never published any of his great advances in mathematics. Some of his ideas did become known through letters, which brings us to Marin Mersenne [French, 1588–1648], a member of a little-known Catholic order called the Mimims who live austerely and often follow intellectual pursuits. Mersenne, through his correspondence with Fermat, Descartes, and the other leading mathematicians and scientists of the seventeenth century—including Galileo and Evangelista Torricelli—acted as a clearinghouse for mathematical ideas, as journals do today. His collected correspondence fills sixteen volumes and discusses physics and music as well as mathematics.

Mersenne's name is most closely associated with a set of prime numbers that he discussed in a 1644 book, *Cogitata physico mathematica*, and at various places in his voluminous correspondence. These are numbers of the form $2^p - 1$, where p is a prime. But not all numbers of this form are primes, and the investigation to find which are and which are not has led to the discovery of many of the largest prime numbers. A Mersenne prime is often shown as M_p, where p is the prime in $2^p - 1$, so, for example, $M_2 = 2^2 - 1 = 3$ and $M_3 = 7$.

The first seven Mersenne primes are 3, 7, 31, 127, 8191, 131,071, and 524,287. For many years the largest known prime was $2^{127} - 1$, a Mersenne prime equal to

$$170,141,183,460,469,231,731,$$
$$687,303,715,884,105,727$$

This number, discovered in 1876 by the French mathematician Edouard Lucas, held the record until the computer age dawned. In 1952 a computer began to spew out higher Mersenne primes and found five of them that year. Subsequently, the largest known prime number has increased dramatically every year or so. Most of the largest known primes are also Mersenne primes, since Lucas and later others developed special methods for testing whether or not Mersenne numbers are prime, but these methods do not apply to primes in general. More recently, newer algorithms have been used by thousands of professional and amateur mathematicians working over the Internet. As of 1999 this group has discovered the largest known prime, which is of course a Mersenne prime, $2^{6,972,593} - 1$, a number with 2,098,960 digits (which I will not provide, especially since this "largest prime" is almost certain to be superseded shortly, perhaps before this book is published). Euclid more than 2000 years ago proved that there is no actual "largest prime"—there are always larger primes.

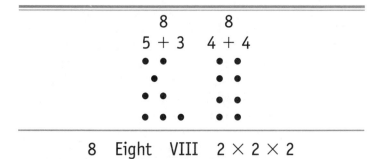

$$8 \quad \text{Eight} \quad \text{VIII} \quad 2 \times 2 \times 2$$

French *huit;* **German** *acht;* **Spanish** *ocho;* **Italian** *otto;* **Latin** *octoni;* **Greek** *octó;* **Papuan body counting,** right shoulder;

classical and medieval finger numerals, left little and ring fingers folded at knuckles, thumb raised

The number 8 occurs as the number of days in a week in German since nights rather than days are counted—so a German week (*acht Tage*) is the eight days that surround seven nights.

FIELD MARKS   Although, technically, 1 is the first perfect cube, the peculiarity of all powers of 1 makes this property not especially interesting. Thus, the next perfect cube— which is of course 8, $2 \times 2 \times 2$—might be called the smallest interesting one. And interestingly, 8 is also the only cube that is 1 less than a perfect square ($3^2 - 1 = 8$).

Other recognizable traits: 8 is the only Fibonacci number other than 1 that is a perfect cube (*see* **Fibonacci and Lucas Families,** pp. 120–122) and a member of the fourth smallest Pythagorean triple, 8, 15, 17.

SIMILAR SPECIES   Other perfect cubes include 27, 64, 125, 216, 343, 512, 729, and 1000.

PERSONALITY   After 4, multiples of 8 are the easiest to recognize. Since 8 is factored into $2 \times 2 \times 2$, it evenly divides into multiples of 10 with any number that also contains a factor of 4; in particular with $4 \times 10 = 40$. Thus, by the time you reach 40 when counting by 8s, the pattern of numerals is established and will thereafter repeat:

$$8, 16, \quad 24, \quad 32, \quad 40,$$
$$48, 56, \quad 64, \quad 72, \quad 80,$$
$$88, 96, 104, 112, 120,$$

and so forth. When the cycle reaches 1000—that is, 1008, 1016, 1024, 1032, 1040, and so forth—it is easy to see that the whole pattern repeats from the beginning. As a result, you need only look at the last three digits of any counting number in the 1000s or greater to observe whether or not it is a multiple of 8. For example, 3124 is not a multiple of 8 (since 124 is not), but 5152 is (since $152 = 8 \times 19$). As this example shows, it may be necessary to test the last three digits by division, because most people do not memorize the multiples of 8 beyond $10 \times 8 = 80$ (although it is certainly useful to memorize $12 \times 8 = 96$, a product that comes up quite often).

ASSOCIATIONS In geometry, 8 appears at the corners of the cube. It is less obvious that as a result of this appearance of 8, it is also the number of faces of the octahedron. If you examine a cube and an octahedron side by side, it becomes clear that the easy way to form an octahedron is to use the centers of the faces of the cube as the vertices of the octahedron. Then cut off the corners of the cube, and triangular faces will form an octahedron. Since there are eight corners to cut off, there will be eight triangular faces.

Just as four colors are needed to color a map on a plane or a sphere and seven to color a map on a torus, eight are needed for all possible maps on a figure with two holes (such as a loving cup).

There are eight knots with a minimum of seven crossings. Beyond this, knotty problems become harder and harder to solve (*see* discussions at **3** and **16**.)

When small numbers of metal atoms are allowed to associate in small clusters, they preferentially form groups with specific numbers of atoms. For sodium atoms heated just enough to form small groups, for example, the clusters tend to form with 8, 20, 40, or 58 atoms. Like the similar magic numbers of nucleons in an atomic nucleus (*see* discussion at **2**), these magic numbers of atoms begin with 8 and 20, magic for stable nucleons; but above 20 the two sequences are different. It is not clear exactly why clusters of atoms form with these magic numbers, but the best guess is that it has something to do with the structure of the five Platonic solids (*see* **5** for an illustration), although the exact correlation is yet to be worked out.

Cubic Family

The perfect cubes are the products formed when any whole number is multiplied by itself—that is, used as a factor—three times. The beginning of the sequence of perfect cubes through $10^3 = 1000$ is given in the preceding text on Similar Species. In addition to memorizing those numbers, it is helpful to be able to recognize 1331, 1728, 2197, 2744, 3375, 4096, 4913, 5832, 6859, and 8000 as perfect cubes, or at least to suspect a possible cubic nature when you encounter them.

Cubes figure in one of the most famous anecdotes about mathematics, a story that is particularly relevant to this book. Since we have the story from G. H. Hardy, who was there at the time, there is every reason to believe it is true.

When the great Indian mathematician Ramanujan, for whom every positive integer was a personal friend, was lying ill, his friend Hardy visited him often. Hoping to cheer him up one day, Hardy tried to think of some interesting mathematical idea to occupy Ramanujan's mind, but nothing came to him. Hardy had, however, noted the number of the taxi he took to the hospital and passed the number on to Ramanujan, but apologized that the number, 1729, was so dull. "No," said Ramanujan, "it is a very interesting number. It is the smallest number expressible as a sum of two cubes in two different ways." (They are $1^3 + 12^3$ and $9^3 + 10^3$.)

The cubes appear in many of the most famous results of number theory, although it is not always clear what spot they occupy. For example, Fermat's Last Theorem, that there is no true statement of the form $a^n + b^n = c^n$ for nonzero a, b, and c and for $n > 2$, can easily be restated as "there is no true statement for cubes or for powers higher than cubes." Or even more dramatically, and certainly misleadingly, as "cubes are the least powers for which there is no true statement of the form $a^n + b^n = c^n$ for nonzero a, b, and c."

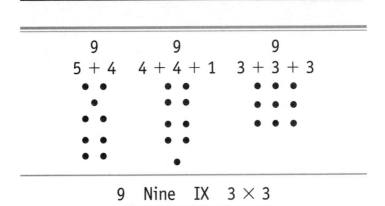

9   Nine   IX   3 × 3

French *neuf*; **German** *neun*; **Spanish** *nueve*; **Italian** *nova*; **Latin** *novem*; **Greek** *ennéa*; **Papuan body counting,** right ear; **classical and medieval finger numerals,** left little, ring, and middle fingers folded at knuckles, index finger pointing, thumb raised

Whether they have retained the information from science classes or crossword puzzles, most people accept that there are nine planets in the solar system—Mercury, Venus, Earth, Mars, Jupiter, Saturn, Uranus, Neptune, and Pluto. Of course, it would seem that the reason there should be nine rather than some other number is a physical fact, but in this case it could be argued that the number is arbitrary, since it depends on what we want to call a planet. When the asteroid Ceres was discovered in 1801 it was considered a planet, for example, and was demoted only when it became clear that there are thousands of similar bodies. Pluto is still treated as a planet, although considerable evidence suggests that it is more like a large comet. Furthermore, we now know that the four inner planets are not much like the giant gas planets of Jupiter, Saturn, Uranus, and Neptune. Perhaps we should speak of the 4 + 4 planets. Rather like a sport utility vehicle.

Pythagorean philosophy, which is based on number, argued that there should be ten planets because 10 is a better number than 9. The Pythagoreans counted the five then-observed planets of the heavens (Uranus had not been noticed, and Neptune and Pluto cannot be seen without telescopes), added in Earth, Moon, and Sun as planets, and so reached a total of 8. They also thought all these bodies revolved about a Central Fire, which would make nine objects. Since they thought a total of ten would be more suitable, they envisioned an unseen planet called the anti-Earth that was presumed to be involved in eclipses.

FIELD MARKS Oddly for a number that has so many uses that derive from its place in the decimal system as 1 less than the base, 9 is seldom encountered when pure numbers are the subject. It is not in any of the famous sequences, with one exception: 9 is a square number.

A set of problems that often occurs in recreational mathematics concerns separating a figure into nonoverlapping subfigures of a particular nature that will exactly cover the larger figure. If the subfigures are squares with counting-number sides and the larger figure is a rectangle, the least possible configuration that forms the rectangle has nine squares. The lengths of the sides of the squares are 1, 4, 7, 8, 9, 10, 14, 15, and 18 units. If you make such squares as cardboard cutouts, you can while away several hours trying to produce the rectangle.

The number 9 appears in connection with the Heegner numbers of fairly advanced mathematics, which are named for the German physicist and mathematician Kurt Heegner [1893–1965]. It had been conjectured that there are only nine of these numbers, specifically −1, −2, −3, −7, −11, −19, −43, −67, and −163. They are related to prime complex numbers (*see* discussion at **Genus *Complex,* 1 + *i***). It was not until 1952 that Heegner proved that this was the complete list, and it was almost another twenty years before mathematicians generally accepted that Heegner's proof was valid.

Also, 9 is a member of the fifth smallest Pythagorean triple, which is 9, 40, 41.

SIMILAR SPECIES The other numbers that are 1 less than a power of 10, such as 99; 999; 9999; 99,999; 999,999; and so forth are most like 9 in the way they are handled in computation, but in fact any natural number that ends in a 9 or a string of 9s shares some of 9's computational utility.

PERSONALITY As noted, the special character of 9 in our system comes from its relation to 10. This relationship influences computation with 9 in several ways as expressed in the decimal numeration system. But these characteristics depend entirely on the numeration system. If the numeration system is changed, the relationship of 9 to 3 or to multiples of 3 does not change, but otherwise 9 becomes just another number.

One of the main tools of mathematics from ancient times through the nineteenth century was "casting out nines," an algorithm that depended on extensions of the divisibility rule for 9: that a number is divisible by 9 if the sum of its digits is also divisible by 9 (*see* discussion at **3**). For example, 123 is not divisible by 9 but 126 is—1 + 2 + 3 = 6, while 1 + 2 + 6 = 9. The proof relies on the fact that the difference of two counting numbers that are divisible by the same number is also divisible by that number (for example, 126 − 27 = 99). A similar argument shows that the sum, product, and quotient are also divisible by that number. Now most numbers are *not* divisible by 9—although half the counting numbers are divisible by 2, and 20% of the counting numbers are divisible by 5, a little thought shows that eight out of nine counting numbers in a finite set, or about 88.8% of them, are *not* divisible by 9.

So if you add, subtract, multiply, or divide two numbers, the chances are good that neither number is divisible by 9 and so the answer will not be either.

The same idea can be extended to remainders, since if two numbers have the same remainder after division by a given number, their sum, difference, or product will also have the sum, difference, or product of the remainders when divided by that number (reduced by 9 if greater than 9). If the remainder is 0, the case is the same as in our example of 126 − 27 = 99. Here is an example with a remainder: 139 has a remainder of 4 when divided by 9 and 21 has a remainder of 3. So the sum 139 + 21 = 160 has a remainder of 4 + 3 = 7, the difference 139 − 21 = 118 has a remainder of $4 - 3 = 1$, the product $139 \times 21 = 2919$ has a remainder of $4 \times 3 = 12$, reduced to $12 - 9 = 3$. Quotients do not work the same way, since, for example, neither $139 \div 21$ nor $4 \div 3$ is a whole number.

This relationship for remainders combined with the simple divisibility rule for 9 is the basis of casting out nines. Here is the way you would use casting out nines to check a product such as $125 \times 394 = 49{,}250$. Begin with the answer. The remainder on dividing 49,250 by 9 can easily be found. It is the sum of the digits, minus 9 re-summed until less than 9. The sum of the digits of 49,250 is $4 + 9 + 2 + 5 + 0 = 20$, which re-summed is $2 + 0 = 2$. So the remainder on dividing by 9 will be 2. Now look at the factors. The remainder for 125 when divided by 9 will be $1 + 2 + 5 = 8$, while that for 394 will be $3 + 9 + 4 = 16$ or $1 + 6 = 7$. The rule for casting out nines is that the product of the remainders will have the same remainder as the product of the original numbers when the answer is correct. So consider 8×7, the product of the remainders, which is 56. Its remainder is $5 + 6 = 11$, which re-sums to 2. So the answer is not shown to be wrong and there is also a good chance (about 88.8%) that the answer is correct—although a simple transposition of digits will go undetected since the sums will not change.

Casting out nines can be explained in terms of an arithmetic in which every number is converted to its remainder upon dividing by some fixed number. This method, an important tool of number theory, is formally called *congruence for numbers* and is also known as *modular arithmetic*. The modular operations all produce results corresponding to the ordinary operations. Here are some examples with modulo 9, although a different modulus would also work. In modulo 9, 37 is converted to 1 because $37 \div 9 = 4$ r 1. Similarly, 49 is converted to 4 because $49 \div 9 = 5$ r 4. Then $37 + 49$ will be a sum that is convertible to $1 + 4 = 5$ in modulo 9.

Checking shows that $37 + 49 = 86$ and that $86 \div 9 = 9$ r 5. Similarly, since the modular numbers $1 \times 4 = 4$, for the regular numbers $37 \times 49 = 1813$ and $1813 \div 9 = 201$ r 4 the predicted relationship again holds true. Modular arithmetic in the guise of number congruence is one of the main tools in the study of the properties of the natural number genus.

The proof of the divisibility test for 9 implies that subtracting the sum of its digits from any counting number will result in a number divisible by 9, whether or not the original number could be divided by 9. Consider 137 for example. The digits sum to 11, so according to this idea $137 - 11 = 126$ should be divisible by 9, and it is. You can also sum the digits more than once to obtain a number that can be subtracted to obtain a multiple of 9. For example, the second sum from 137 is $1 + 1 = 2$ and $137 - 2 = 135$, another number divisible by 9.

Here is another divisibility relation for 9 that always works. If you reverse the digits of any number and subtract the smaller from the former, the result will be divisible by 9. To consider 137 once more, reversed it becomes 731. Then $731 - 137 = 594$, which is divisible by 9. This is similar to a proof that adding the digits reveals divisibility by 9 or by 3, this property can be observed if you write the original numeral $abcd$ (to use a four-digit example) as the number $d + 10c + 100b + 1000a$. Reversed it is $a + 10b + 100c + 1000d$. Subtracting the first from the second gives $1000d - d + 100c - 10c + 10b - 100b + a - 1000a = 999d + 90c - 90b - 999a = 9(111d + 10c - 10b - 111a)$, which is clearly a multiple of 9. The only reason for subtracting the larger from the smaller is to give a counting number instead of a negative answer.

ASSOCIATIONS The nine-point circle was one of the rare discoveries of elementary plane geometry missed by the ancient Greeks. It lay unnoticed until about 1820, when it was found independently in France and in Germany. The nine points of the theorem are special points of a triangle that come in three sets. If a circle passes through the three points of one of the sets, it also passes through the three points of the other two, for the total of nine points.

The first such set (D, E, F in the illustration below) can be thought of as the place where the altitudes from each vertex meet the opposite sides of the triangle. The second set (A', B', C') is the midpoints of the sides of the triangle. The third set of points is harder to describe. The three altitudes meet in a point, H; connect that point to each of the vertices of the triangle (ABC) and find the midpoints of the connections. Those midpoints (A'', B'', C'') are the third set of 3.

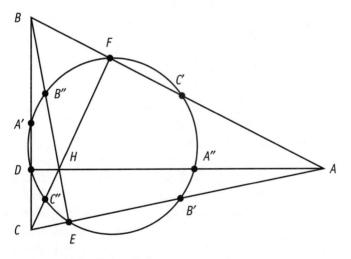

Nine-Point Circle

I have been unable to find any source of the myth that a cat has nine lives (it certainly has nothing to do with the whip called a "cat o' 9 tails"). I suspect that the nine lives come somehow from Dante's nine circles of Hell. In the poet's medieval Christian world cats were viewed as symbols of the old pagan religion and often treated quite badly, but since they are hardy, agile, and elusive animals, they often survive or escape events that would extinguish other creatures.

It would appear that rhesus monkeys are capable of counting to 9. In 1998 two psychologists from Columbia University were able to demonstrate that each of a pair of rhesus monkeys could arrange nine objects in a particular order. The monkeys, Rosencrantz and Macduff, had been trained to be able to arrange four fields each with a different number of objects from 1 to 4 in order. After eleven sessions with these fields (presented to the monkeys on a touch-screen computer), Rosencrantz and Macduff were set to work on arranging nine fields with numbers of objects from 1 to 9 in order. Rosencrantz managed to be right almost all the time and Macduff was only slightly behind. The trials do not seem to have extended beyond 9, so it is not clear how far Rosencrantz and Macduff could have gone.

Although Euclid's proof that there are only five regular, or Platonic, solids (*see* discussion at **5**) is treated as the crowning achievement of the *Elements*, Johannes Kepler [German, 1571–1630] was skeptical. He showed in 1619 that two star-like polyhedra (with several projecting points) fit Euclid's criteria for regular solids, so there are at least seven regular solids. Almost 200 years later the little-known French mathematician L. Poinsot independently rediscovered Kepler's starlike polyhedra and added two more of his own. Although there are those who in various ways quibble with the definition of "regular" (some trying to make the traditional view of five regular polyhedra correct and others expanding far beyond the regular polyhedra of Euclid–Kepler–Poinsot), a conservative statement of the situation would be that there are nine regular polyhedra—the five Platonic solids and the four starlike solids.

The earliest magic square was based on 9. This one is a Chinese invention from the fourth or fifth century B.C. known as the *lo shu*. Legend has it that it was first seen by the legendary King Yu on the back of a turtle in the River Lo around the twenty-third century B.C., but there is no confirming evidence of it that early.)

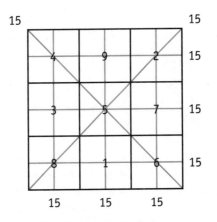

The Earliest Magic Square

As you can see, the "magic" part of the *lo shu* is that each row, column, and both diagonals add to the same sum, 15. The *lo shu* is classified as a magic square of order 3 because it has three rows and three columns. Not only is this the first magic square, it is also the only magic square of order 3 (ignoring the symmetric rotations and reflections that produce essentially the same square). Since there are no magic squares of order 2, the *lo shu* is also the smallest possible magic square.

Herb gardeners say that parsley seed must go to the Devil nine times before it begins to grow as a plant. Parsley does take uncommonly long to germinate, and an impatient gardener might well think it has gone to the Devil!

In the Roman Catholic Church, a novena is nine days of prayer and meditation.

A firkin, equal to nine imperial gallons, is an old British measure for beer or ale.

---

10   Ten   X   2 × 5

French *dix*; **German** *zehn*; **Spanish** *diez*; **Italian** *dieci*; **Latin** *decem*; **Greek** *déka*; **Papuan body counting,** right eye; **classical and medieval finger numerals,** left index finger touching last joint of thumb

Linguists have traced the origin of the number words we use to an original and now lost language called Proto-Indo-European, which is thought by some to have been spoken about 3000 B.C. Proto-Indo-European is the reconstructed source of the Germanic languages (including English) and the Romance languages derived from Latin, as well as of medieval and classical Latin and ancient Greek, Sanskrit and the modern languages of India and Persia, a large group of Turkish languages, Russian and the Slavic group, and Celtic—in short, most, but not quite all, of the ancient and modern languages of Europe and eastern Asia. One of the key methods used in reconstruction has been to look for words common to all these languages, words that include the key number words. Of these, the Proto-Indo-European word for "ten" is thought to have simply meant "two hands." The word for "hundred" was just "ten times." There was no Proto-Indo-European word for "thousand" because Proto-Indo-Europeans could not count that high (say the linguists).

FIELD MARKS The number 10 is the only whole number that has a factorial (*see* **Factorial Family,** pp. 145–146) that is nontrivially the product of factorials: $10! = 7!6!$. While it is always possible to string together a pair of factorials to create a product that is a factorial, the general method is treated as trivial since an infinite number can be generated by what mathematicians consider the essentially trivial formula $k!! = (k! - 1)!k!$. For $k = 1$ this is $1!! = 0!1!$ or $1 = 1$; for $k = 2$ this is $2!! = 1!2!$ or $2 = 2$; if $k = 3$ it is $3!! = (3! - 1)!3!$, or $6! = 5!3!$, which is

$$6 \times 5 \times 4 \times 3 \times 2 \times 1 = (5 \times 4 \times 3 \times 2 \times 1) \times (3 \times 2 \times 1)$$

Since 6! is the same as 3!!, a trivial solution can be combined with the single nontrivial one cited above to write $10! = 7!5!3!$, but this is not the only nontrivial instance of a factorial that can be written as the product of exactly three other factorials (try 16! on your own).

A famous unproved idea of mathematics is the conjecture of Christian Goldbach [German, 1690–1764] that all even numbers 4 or greater can be expressed as the sum of two primes (usually stated in the oddly less powerful form that all even numbers greater than 4 can be expressed as the sum of two odd primes). A sidelight to Goldbach's conjecture is that 10 is the smallest even number that can be expressed as the sum of two primes in two different ways: $10 = 5 + 5 = 3 + 7$.

Other recognizable traits: 10 is a triangular number ($4^\triangle = 10$).

SIMILAR SPECIES   The powers of 10, starting $10^0 = 1$, $10^1 = 10$, $10^2 = 100$, $10^3 = 1000$, and $10^4 = 10,000$ and continuing thus are all formed by 1 followed by 0s to the number in the exponent. These powers are among the most familiar and useful of all numbers and the basis of the decimal system. When written as decimal fractions, the negative powers can also be explained in terms of counting 0s, provided the 0 in front of the decimal point is always used and counted. Thus, $10^{-1} = 0.1$, $10^{-2} = 0.01$, and $10^{-3} = 0.001$.

PERSONALITY   The number 10 and its powers are the basis of nearly all computational strategies, including the standard algorithms as well as techniques of mental computation. When in a standard algorithm you rename or, as teachers long ago said, "carry," a number in a problem such as $25 + 27$, you may think "carry 1" as you take the first digit from the sum 12 of $5 + 7$, but of course that 1 represents a 10. The multiplication algorithm not only uses a similar kind of renaming but also relies on the property in decimal numeration that multiplying by a power of 10 moves each decimal digit one place to the left.

In mental calculation, the proximity of a number to a multiple of a power of 10 is a welcome help. For example, to multiply any whole number by 799, think 799 is 1 less than 800 (which is 8 times 10 squared), so I can multiply by 8, annex two zeros to multiply that result by 100, and then subtract

102

the other factor to take away the extra 1 that I used for 800. Thus, 15×799 is $(15 \times 800) - 15$, or $12,000 - 15 = 11,985$.

ASSOCIATIONS Throughout scientific and mathematical writing or in science or math classrooms we encounter the phrase "order of magnitude" or "orders of magnitude." Oddly, I have never come across a teacher or a textbook that explained what the expression means. Dictionaries reveal the meaning, but you have to look under "order," not "magnitude." Several of them suggest that an order of magnitude is a range from a given value to ten times that value, but in most instances it is used to indicate a factor that is a power of 10. For example, if someone says that an ant can travel an order of magnitude faster than a snail can, the meaning is that the ant's speed is about ten times that of the snail. The idea of a range appears here since we have not specified either speed, so if the range of the snail's speed is taken to extend from 0 to 1 mph, for example, then the ant, at an order of magnitude greater, travels from 1 to 10 mph. For the plural form, several multiples of 10 are meant, so that the mass of Earth is many orders of magnitude greater than that of a person.

The Greek numerologists, such as the Pythagoreans, were exceedingly fond of 10, even postulating an extra heavenly body to make the total number of "planets" come to ten (*see* discussion at **9**). One of the Pythagoreans, Philolaus of Tarentum [Hellenic, *c.* 425 B.C.], claimed that 10 is "great, complete, all-achieving, and the origin of divine and human life and its Leader."

In the twentieth century we came to accept that the universe, once thought to be fully described by three dimensions of space, is better understood as having four dimensions (with time as the fourth dimension). Since the 1920s, however, several related theories have proposed even more radical revisions of dimensionality, ideas that have come to be taken very seriously in the last quarter century. One of the most widely accepted—superstring theory—postulates that the universe is ten-dimensional. In superstring theory, the fundamental entities from which everything is built are not pointlike particles but very short curves, or strings, with one dimension—length. The strings are typically as short compared with an atom as an atom is compared with the solar system. The strings exist in ten dimensions, including the three familiar spatial dimensions of length, width, and height, one time dimension, and six other spatial dimensions that must be wound up so tightly that we do not observe them. (There is still more to the dimension story; *see* discussions at **11**.)

The number 10 is a magic number in chemistry because atoms with eight electrons in their outer shell and two electrons in the first shell are stable (the element with this configuration is neon). Furthermore, atoms that have numbers of electrons close to this magic number will combine chemically to produce a filled second shell of eight electrons, so that each atom can pretend that it has the magic ten electrons.

It appears that one form of pond scum can count to 10. The blue-green alga *Anabaena* forms strings of cells with nine normal, photosynthesizing cells in a row and then a tenth that takes nitrogen from the air and converts it to a form that plants and algae can use. It was shown in 1998 that the gene *patS* encodes an inhibitor in a cell that will develop into a nitrogen-fixer. This signal prevents nearby cells from becoming nitrogen-fixers themselves, but it weakens over the distance occupied by about nine cells. Then ten cells farther along the chain of cells, the suppression becomes too weak to be effective and a new nitrogen-fixer develops.

Anabaena is not a very good counter, however, because it uses analog means to a digital end. Often there are eight cells, or eleven or twelve, instead of nine between nitrogen-fixers. Although modern technology has triumphed by using digital means to achieve analog ends, as in a compact disc (CD) or digital video disc (DVD), or a computer hard drive, the opposite route is much less accurate. Thus, pond scum may not be so smart after all.

Christians and Jews associate 10 with the Ten Commandments that Moses received from God on Mount Sinai. Oddly, however, there is no uniform agreement on how to group and number them to obtain the total ten, although their name comes from Moses himself. Here is a brief summary in the most obvious numbering:

1. No other gods before me.
2. No graven images.
3. Don't take name of Lord in vain.
4. Keep the Sabbath.
5. Honor your parents.
6. Don't kill.
7. Don't commit adultery.
8. Don't steal.
9. Don't bear false witness.
10. Don't covet.

Decimal System

The basic scheme of the Hindu-Arabic decimal system is familiar to everyone—the ten digits 0, 1, 2, 3, 4, 5, 6, 7, 8, 9 and place value based on powers of ten (here spelled out to make the point)—although perhaps like a good marriage, it is so familiar in its workings that we fail to think about how special and wonderful it is. In a *place-value* system, the position in which a digit appears in a numeral determines the amount that the digit is worth to the number. The places to put the digits begin on the right of the numeral and move toward the left, increasing by a power of ten with each place. This right-to-left progression is part of the heritage of the Arabs, who are accustomed to it from their writing.

Every place-value system needs a symbol for an empty place. This need was perceived only gradually in the history of place-value numeration systems, but it eventually came to be recognized in all cultures that used such systems. The concept came to be called a *place-holder* by students of mathematics and is fulfilled in our system by 0. The idea of using 0 as a place-holder is often kept separate from the idea of 0 as a number on its own, which developed later. But from a modern mathematician's point of view, 0 is, with 1, the most fundamental of numbers, quite essential in any scheme of numbers that goes beyond the two original mathematical operations of counting and measuring. Since 0 is not involved in an obvious way in counting or measurement, early views of numeration could comfortably place 0 in the numeral category. When the calculus and related techniques for dealing with infinite sequences and series were invented in the seventeenth century, many new roles for 0 appeared, but even in these roles it was not clear that 0 was actually a number. The seventeenth-century appearances of 0 were rather like the concept of *nothing*, which is a related but different idea—the difference is similar to that between 0 and the empty set. It was not until the nineteenth-century development of numbers as systems with formal structures that 0 was recognized fully as a number. At that point it also became apparent that the place-holder idea, which used 0 only as numeral, could be replaced by treating 0 exactly the same as the other nine digits, thus letting it fill both roles.

The Hindu-Arabic system, like nearly all powerful modes of thought, manages to keep two entirely separate concepts under the same roof (F. Scott Fitzgerald proposed that the ability to hold contradictory ideas in the mind simultaneously, and still be able to act, is the test of a first-rate intelligence). The digits are what appear on the page, so they are numerals. But they are also the multipliers for the powers of ten, so they are numbers, since you cannot multiply by a symbol. Acceptance of paradox instead of struggle against it is often the beginning of wisdom. Exponents, tiny numerals placed as superscripts to the right of other numerals, also should be treated as numbers as well as numerals.

Since ten to the 0 power is one and ten to the power of 1 is ten, the Hindu-Arabic system begins with the power of 0 as the rightmost place. Thus, the number written as 10, which is the smallest to use place value, means 1 ten and 0 ones. Another way to describe 10 is as $(1 \times 10^1) + (0 \times 10^0)$. This is a bit circular, because the symbol for ten is given as 10, but it is a basis from which to work. Once you accept 10 as a way of writing ten, then any number of any size, even those we call decimal fractions or simply decimals, can be written in this expanded form that shows more of the bones of the system. It is necessary to use negative integers as exponents for decimal fractions that are not whole numbers, but they follow the scheme as well. A number shown as 523.67 is interpreted as $(5 \times 10^2) + (2 \times 10^1) + (3 \times 10^0) + (6 \times 10^{-1}) + (7 \times$

10^{-2}). The same procedure can be extended for an infinite number of places in either direction. Numbers written in the Hindu-Arabic system show certain properties that derive entirely from the method used in writing the numeral, while other properties of numbers come from the numbers themselves. It is not always obvious whether a property comes from the method used to name it or is intrinsic to the number. When we observe that all whole-number multiples of ten end in 0, for example, we are at best subliminally aware that this is true only in the Hindu-Arabic numeration system. In another system, based on nine or twenty, for example, this property would vanish. For example, a place-value system similar to the Hindu-Arabic in all ways but with "nine" replacing "ten" would use 10 to mean 1 nine and 0 ones, while 11 would mean 1 nine and 1 one, or ten. Such a system would have the property that all nine digits—0, 1, 2, 3, 4, 5, 6, 7, 8 with no need for 9—appear as the last digits of whole-number multiples of ten. The effort to bring this concept to the forefront of the mind was another rock upon which the "new math" crumbled, so I mention it with some trepidation.

An intrinsic property of ten would be that it can be factored as "two times five" no matter what system is used for writing the numbers. Thus, if nine is used as a base, then the symbols $2 \times 5 = 11$ represent a true statement. Properties such as being even or odd, positive or negative, rational or irrational, all belong to the numbers themselves, not to the ways they are written.

11   Eleven   XI   prime

French *onze*; **German** *elf*; **Spanish** *once*; **Italian** *undici*; **Latin** *undecim*; **Greek** *héndeka*; **Papuan body counting,** left eye; **classical and medieval finger numerals:** left index finger touching last joint of thumb (sign for "ten") with little finger folded at joints (sign for "one")

According to linguists, in Proto-Indo-European, the ancestor language to English and German, the original word for eleven meant "one left over," which is preserved in the words "eleven" and "elf." But in the Romance languages, that idea has been changed into a word similar to the Latin *undecim*, which means "one [more than] ten."

FIELD MARKS The number 11 when written in the Hindu-Arabic decimal system is a rep-unit, a number that repeats the single digit 1; it is also a prime.

The number 11 is also the number of ways that six indistinguishable objects can be arranged into separate sets (these are //////, ///// /, //// //, //// / /, /// ///, /// // /, /// / / /, // // //, // // / /, // / / / /, / / / / / /), so it is known as a *partition number*. The partition numbers are similar in their origin to the Bell numbers (*see* discussion at **52**), which are the number of ways that distinct objects can be arranged in separate sets. While the partition number for 6 is 11, the Bell number is 203, because for distinct objects it is possible to tell one arrangement of, say, three objects, two objects, and one object from another that has the same configuration but with the objects placed in different sets (AB CDEF is distinguishable from BF ACDE). The exact formula that yields the partition number for a single set is an immensely complicated equation that involves π, $\sqrt{2}$, a function based on e (**see Genus *Real*, 2.71821 . . .**) called the hyperbolic sine, and the number 24.

Other recognizable traits: 11 is a Lucas number (*see* **Fibonacci and Lucas Families,** pp. 120–122), a prime twin with 13, and a member of the Pythagorean triple 11, 60, 61.

SIMILAR SPECIES The next rep-unit is 111, but in the decimal system this number is easily recognized as divisible by 3. There seems to be no general rule that anyone has found for knowing if a number formed by repeating the digit 1 is prime or not. It is relatively easy to test small numbers. You must, however, go up to 1,111,111,111,111,111,111 (1 used 19 times) before finding another prime. But then the next one is just a few repetitions away— 11,111,111,111,111,111,111,111, which is 1 used 23 times. Thus encouraged, you might look for the next rep-unit at perhaps 29 or 31 uses of 1 (noting that 2, 19, and 23 are all primes themselves). In fact, the next prime rep-unit has 1 used 317 times. Testing a few factors shows that 317 is also prime. How many rep-units are there? No one knows. The number could be infinite.

Note that using a different digit (with one possible exception) cannot produce a similar sequence of primes in our numeration system—the even digits 0, 2, 4, 6, and 8 cannot be used repeatedly to form primes except for the trivial case of 2 used once. The divisibility rules for 3, 5, and 9 exclude all numbers of the form 333 . . . , 555 . . . , and 999 . . . , except again for the trivial 3 or 5 standing alone. This leaves 7 to consider. After the trivial 7 alone, you find $77 = 7 \times 11$, then 777 divisible by 3, and so on. The lowest number that a quick check revealed to be a possible rep-seven prime was 7,777,777, with 7 used 7 times. At this point the checking becomes more difficult, although I eventually located $239 \times 32,543$ as factors of this rep-seven. The problem of whether or not there are nontrivial prime rep-sevens has not to my knowledge been resolved.

The other small partition numbers are 1 (0 elements), 1 (one element), 2 (two elements), 3 (three elements), 5 (four elements), 7 (five elements), 15 (seven elements), 22 (eight elements), 30 (nine elements) 42 (ten elements), and 56 (eleven elements).

PERSONALITY The basic facts for the elevens times table are so simple that anyone who can count does not even need to learn them, as the sequence of products is 11, 22, 33, 44, 55, 66, 77, 88, 99, and 110. Add $11^2 = 121$ to the list and you have a handy group of products with no visible effort on your part.

Computation with 11, especially multiplication, is made easier using the Latin or Italian idea of "one more than ten" to mean 11. For example, to multiply 496×11, think $4960 + 496 = 5456$. For once, the actual computation is the same as the standard algorithm, since most people usually write

$$
\begin{array}{r}
496 \\
\times\, 11 \\
\hline
496 \\
496 \\
\hline
5456
\end{array}
$$

For large numbers, however, it is easier to multiply by 11 (or similar numbers 111 or 1111 or the like) by simply writing the number more than once with an offset to the left for each succeeding numeral and adding.

As you might expect from the divisibility test for 9 (*see* discussions at **3** and **9**), or 1 less than 10, there is a digit-related divisibility test for 11 that also stems from proximity to 10. Start at the right end of a large number and, moving to the left, subtract the second digit from the first, then add the third to your previous result, subtract the fourth, and so forth until you reach the last. If the result is divisible by 11 (counting 0 as divisible by 11), then so is the original number. Consider, for example, 2,649,647. Think $7 - 4 = 3 + 6 = 9 - 9 = 0 + 4 = 4 - 6 = -2 + 2 = 0$, which means that the original is also divisible by 11 (2,649,647 ÷ 11 = 240,877).

ASSOCIATIONS The number of dimensions in the real universe, after being accepted as three for ages, has kept growing throughout the twentieth century, with four dimensions explicitly stated in 1907, five dimensions made plausible in 1926, and ten dimensions in the most common superstring theory of 1984. Since 1994 there has been a period of dramatic progress. What once appeared to be five different superstring theories are actually different realizations of a unique underlying concept, which has been named M theory (for membrane theory, although the theory has gone beyond the idea of mathematical models of thin, flexible sheets). Many physicists believe that M theory is a conceptual revolution as profound as those associated with relativity and quantum theory. M theory, as well as some versions of superstring theory, is generally cast in terms of a universe of eleven dimensions, with seven dimensions curled up so tightly as to be unobservable (but other numbers of dimensions are possible).

The smallest Mersenne number that is composite instead of prime is $2^{11} - 1$ or 2047, which equals 23×89.

---

12   Twelve   dozen   XII $2 \times 2 \times 3$

French *douze*; **German** *zwölf*; **Spanish** *doce*; **Italian** *dodici*; **Latin** *duodecim*; **Greek** *dódeka*; **Papuan body counting,** nose; **classical and medieval finger numerals,** left index finger touching last joint of thumb (sign for "ten") with little and ring fingers folded at joints (sign for "two")

In the Proto-Indo-European ancestor language of English, the original word for 12, similar to that for 11, meant "two over." This meaning is carried into German as well as English, although replaced with "two [more than] ten" in Romance languages. In Germanic tradition, 12 is the basic unit of measurement, as in 12 pence to the shilling, 12 troy ounces in a pound of gold, and 12 inches in a foot. This was true in early times, as can be observed in Old Norse, which even uses "12 and 3" in some instances to mean 15. The primacy of 12 was firmly established in much of Europe by Charlemagne in the eighth century.

FIELD MARKS   An *abundant number* is a whole number for which the sum of its divisors, not including the number itself, is greater than the number. The lowest abundant number is 12, since its proper divisors (excluding itself) are 1, 2, 3, 4, and 6, which sum to 16. (A deficient number is less than the sum of its divisors.)

If the sum of the proper divisors is interesting, the product must be as well. One way to define a prime would be to say that it is a number whose only proper divisor is 1, so only composite numbers are worth examining for the sums of their proper divisors. One perfect number is equal to the product of its proper divisors (*see* discussion at **6**) as well as the sum. The products of proper divisors of some other small composite numbers are 2 (for 4), 8 (for 8), 3 (for 9), 10 (for 10)—and 144 (for 12), which seems a sudden leap from the others. The number 12 has sometimes been identified as the smallest whole number for which the product of the proper divisors is a perfect square.

Other recognizable traits: 12 is a member of the next-to-least Pythagorean triple of 5, 12, 13 as well as of the less well known triple 12, 35, 37.

SIMILAR SPECIES Some other abundant numbers are multiples of 12, such as 24 (1, 2, 3, 4, 6, 8, and 12 sum to 36) and 36 ($1 + 2 + 3 + 4 + 6 + 9 + 12 + 18 = 55$). But 18 also is an abundant number. The next logical question is: Is being a multiple of 6 required? The answer is no, since the first few abundant numbers include 20 ($1 + 2 + 4 + 5 + 10 = 21$) as well as the multiples of 6. So far the list is all even: 12, 18, 20, 24, 36. Is evenness required? No, since 945 is an odd abundant number ($1 + 3 + 5 + 7 + 9 + 15 + 21 + 27 + 35 + 45 + 63 + 105 + 135 + 189 + 315 = 975$). In fact, 945 is the least odd abundant number.

This suggests the related question for perfect numbers—are any perfect numbers odd? No one knows, although long computer-aided searches have failed to find any (*see also* discussion of *multiply perfect numbers* at **120**).

The Germanic "great hundred" is 120, which is to 100 what 12 is to 10, and the "great thousand" is 1200. The remnant of this old Germanic unit in English is the long ton, which is 2240 pounds instead of the 2000 pounds of the ordinary ton (another name for the great hundred in English is the long hundred). Note that a long ton is 2 measures of a thousand pounds plus 2 measures of a great hundred pounds.

PERSONALITY Multiplication by 12 is aided considerably if you have memorized the times table through 12 in the beginning. The sequence of products 12, 24, 36, 48, 60, 72, 84, 96, 120 is probably familiar and recognizable as having 12 as a factor even if you have not made a conscious effort to memorize it, and $144 = 12 \times 12$ is familiar to most people as well. The only oddball is $132 = 11 \times 12$.

It is again easy to use the nearness of 12 to the friendly 10 for multiplying larger numbers by 12. Combine multiplying by 10 and doubling, two of the simplest operations. For example, to find 438×12, you might think $4380 + 876 = 5256$.

The divisibility test for 12 is a combination of the divisibility rules for 4 and 3. This is easier in practice than you might think. Consider some larger number such as 183,456,972. Is it divisible by 12? Check the last two digits; 72 is divisible by 4. Check the sum of the digits, which is 45, divisible by 3; therefore, 183,456,972 is also divisible by 3. Since it is divisible by both 4 and 3, it is also divisible by 12. Checking shows that $183,456,972 \div 12 = 1,528,808$ with no remainder.

ASSOCIATIONS In geometry the number 12 appears several times in connection with the Platonic solids. For example, both the cube and the octahedron have twelve edges, the icosahedron has twelve vertices, and the dodecahedron has twelve faces—for the same reason that the number of vertices of the cube is the same as the number of faces of the octahedron (*see* illustration at **5**).

The sphere-packing problem for three-dimensional space (*see also* discussion of circle-packing problem at **6**) was conjectured by Kepler in 1611 to be an arrangement in which each of twelve spheres of a given size touches a single sphere of the same size. It was not until 1874 that a proof was given that this was the maximum number of spheres that could touch, and not until 1998 that anyone showed that Kepler's approach was shown to be the best packing.

One result of Euler's Theorem for polyhedra (*see* discussion at **2**) is that every molecule of fullerene, a form of carbon that has its atoms arranged in a pattern based on a geodesic sphere, will contain exactly twelve pentagons with the rest of the carbon atoms arranged in hexagons. This is a specific application, also established by Euler, of the general principle that all closed figures made from pentagons and hexagons must contain exactly twelve pentagons. The smallest such fullerene, known officially as buckminsterfullerene and unofficially as the buckyball, contains 60 carbon atoms, which are arranged into the 12 pentagons ($12 \times 5 = 60$) with 0 hexagons. Carbon nanotubes, fullerenes that can consist of millions of atoms, all contain exactly twelve pentagons no matter how long they are, with the rest of the carbon atoms arranged in hexagons.

According to the Revelation of St. John the Divine there are twelve gates to heaven, one for each of the traditional Twelve Tribes of Israel as well as for the Twelve Apostles.

It is sometimes handy to know that 1,000,000 seconds is about the same as twelve days.

Denominate Numbers

The combination of a number and a measurement unit, such as 12 inches, is sometimes called a *denominate number*, as opposed to a *pure number* such as 12. Denominate numbers are useful tools as well as necessary for the expression of measurement. They help physicists find which operations are suitable for a particular problem and also whether a series of mathematical steps has produced a correct result. All problems in the motion of bodies or the applications of forces, for example, involve denominate numbers that express combinations of three dimensions—length, mass, and time. The measures may be squared or cubed or used as divisors, but they are, at bottom, the same basic units. For example, the denominate number 12 square miles is based on the length *mile* and can be expressed as 12 miles^2. Similarly, an acceleration of 32.2 feet per second per second is expressed as 32.2 ft/sec^2 where the slash (/) indicates *per* as well as division.

The dimensions are abstracted in practice. Length, mass, and time can be expressed as L, M, T and treated as entities in themselves, a method known formally as *dimensional analysis*. A quantity, such as force, is represented as ML/T^2, meaning "mass-length per time squared"; this is the dimensional structure of any force. The actual force may be, for example, 12 newtons, but since a newton is a kilogram-meter per second per second (a mass of 1 kilogram accelerated by a meter per second per second) that is the same as 12 kilogram-meters per second per second. All the main denominate numbers of physics can be similarly expressed: velocity, such as miles per hour, is L/T, work is L^2M/T^2, and angular momentum is L^2M/T.

Every equation in physics can be expressed as dimensions only and, if true, will always have exactly the same dimensions on each side of the equals sign. For example, the familiar equation $F = ma$ for force = mass × acceleration can be analyzed dimensionally as force = ML/T^2, mass = M, and acceleration = L/T^2, so the dimensions are the same: $ML/T^2 = ML/T^2$.

To use this idea in problem solving, you set up the dimensions of the expected answer. For example, if you know the answer must be speed, then it will have the dimension L/T. Then go through the calculations ignoring the numbers and their specific denominations, treating them as combinations of length, mass, and time. The result of these calculations should be in the same dimensions as the expected answer. If not, the nature of the deviation from the expected answer often illustrates the flaw in your reasoning. When the dimensions are correct, it becomes easy to insert actual denominate numbers and derive the answer.

13   Thirteen   XIII   prime

French *treize*; **German** *dreizehn*; **Spanish** *trece*; **Italian** *tredici*; **Latin** *tredecem* or *decem et tres*; **Greek** *treis kaí deka*; **Papuan body counting,** mouth; **classical and medieval finger numerals**, left index finger touching last joint of thumb (sign for "ten") with little, ring, and middle fingers folded at joints (sign for "three")

Why did 13 emerge as an unlucky number in Western culture? The obvious suspects are the association of Christ and twelve apostles (the thirteenth member of the group is Judas) or the counterversion, Satan and a twelve-member witch's coven. Today we avoid adding a thirteenth guest to a dinner table—in a coven the thirteenth guest would always be the Devil, who arrives when the twelve witches are assembled. Interestingly, 13 was considered lucky by the Maya.

FIELD MARKS The famous American mathematical writer and inspiration and source to all of us, Martin Gardner [b. 1914] seems to be the first to have recognized that the numbers of holes in a Chinese checkerboard form a particular pattern of numbers. If we count 1 as a trivial example, then the first Chinese checkerboard number is 13, when shown as an array of dots like the following:

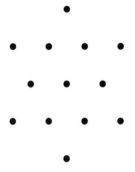

Gardner has named these the star numbers. If the symbol ★ is used to indicate the star numbers, then $1^\star = 1$ and $2^\star = 13$. Gardner's formula for the numbers is $n^\star = 6n(n - 1) + 1$.

Other recognizable traits: 13 is a Fibonacci number (**Fibonacci and Lucas Families,** pp. 120–122); a member of the second and eighth Pythagorean triples (5, 12, 13 and 13, 84, 85); a prime twin with 11; and a Wilson prime (*see* discussion at **5**). Finally, an octagon can be dissected into thirteen nonoverlapping, noncongruent octagons if you are very clever.

SIMILAR SPECIES When written as Hindu-Arabic numerals, the number 13 and its square exhibit some unusual symmetries. The first is that both 13 and 169, the square, are unusual in that the sum of the digits of 13 is a perfect square ($1 + 3 = 4$) and the sum of the digits of 13's square is the square of the earlier sum ($1 + 6 + 9 = 16 = 4^2$). In a similar symmetry the square of the reverse of 13 (that is, 31^2) is the reverse of the square of 13 ($31^2 = 961$, the reverse of 169).

The next few star numbers after 13 are 37, 73, 121, 181, 253, 337, 433, and 541. The number of holes on an actual Chinese checkerboard is 5★ = 121.

PERSONALITY Although 13 is a prime number, its appearance as a quarter of 52, as in 52 weeks of the year or as a suit of cards, makes it more familiar than many other small primes. For example, the multiples 26 and 39 are both so commonly used that you should not have to multiply to find them. For products with larger numbers, breaking 13 into 10 + 3 makes it possible to multiply first by 10 (annexing a 0) and then by 3, which for factors of two or more digits can be done by doubling and then adding in the number.

ASSOCIATIONS In the United States, the number 13 has a large set of associations that all stem from the accident that the nation was formed from thirteen colonies. The eagle on the Great Seal of the United States is clutching thirteen arrows and an olive branch with thirteen leaves and thirteen olives. Above the eagle's head a cloud surrounds a constellation of thirteen pentagrams arranged in the star pattern of thirteen shown above for 2★. The American flag reflects history with its thirteen stripes.

Among mathematics historians and even some mathematicians, the number 13, usually written as XIII, is familiar as the number of books in Euclid's *Elements*.

Fibonacci and Lucas Families

The Fibonacci numbers are any of the counting numbers found in a sequence formed by adding the two preceding numbers of the sequence to find the next, starting with an initial pair of 1s:

$$1$$
$$1$$
$$1 + 1 = 2$$
$$1 + 2 = 3$$
$$2 + 3 = 5$$
$$3 + 5 = 8$$

and so forth, producing the Fibonacci sequence 1, 1, 2, 3, 5, 8, 13, 21, 34, 55, 89, 144, 233, . . .

Mathematicians have found many interesting properties of the Fibonacci numbers and sequence, including applications to understanding growth patterns in plants, but the whole idea originated in the following problem: "How many pairs of rabbits will be produced in a year, beginning with a single pair, if every month each pair bears a new pair that becomes productive from the second month on?" As you may have guessed, the answers for each month are 1 pair, 1 pair, 2 pairs, 3 pairs, 5 pairs, 8 pairs, . . . The problem was among many proposed and solved by the most influential mathematician of the Middle Ages, Leonardo of Pisa, also known as Fibonacci ("son of Bonaccio") [Italian, *c.* 1170–*c.* 1250]. Leonardo also contributed to the development of algebra and helped introduce decimal notation to Europe.

Edouard Lucas [French, 1842–1891], who studied the Fibonacci series in depth, also generalized the concept to any sequence formed by adding the two preceding numbers to obtain the next member of the sequence. The Fibonacci sequence is usually stated as beginning with 1, 1 although the same numbers arise if you begin with 0, 1. You also get the same sequence if you begin with 1, 2. But a new sequence arises when you begin with 1, 3. This generating pair results in 1, 3, 4, 7, 11, 18, 29, 47, 76, 123 as the first ten members, a sequence of numbers called the Lucas numbers in honor of its inventor. Other generalized Fibonacci sequences obtained by starting with some other pair of numbers are simply called "generalized Fibonacci sequences."

There are many unusual appearances of the Fibonacci sequence and its generalizations that have been discovered. One that applies to all the generalizations is that the ratio between two consecutive members of a sequence eventually approaches the Golden Ratio ϕ (*see* discussion at **Genus Real,** 1.61803 . . .). There are many other relationships between ϕ and Fibonacci or Lucas numbers. One of the most surprising is that the nth Lucas number is always ϕ^n rounded to the nearest natural number.

There is a sparseness of squares and cubes among the Fibonacci and Lucas numbers that is astounding considering that both are infinite sets. If you discount the trivial example of 1, the only Fibonacci number that is also a perfect square is 144, and the only perfect square among Lucas numbers is 4. Similarly, the only perfect cube that is a Fibonacci number is 8, and there are no perfect cubes among the Lucas numbers. Just as scarce are the numbers (discounting 1 again) that are both Fibonacci and Lucas numbers: the only example is 3.

There is also a second way to generalize the Fibonacci numbers. Instead of adding two numbers each time, you can add n numbers. The first such sequence, adding three numbers each time and starting with 0, 0, 1, is sometimes called the tribonacci numbers: 1, 1, 2, 4, 7, 13, 24, 44, 81, 149, . . .

14 Fourteen XIV 2 × 7

French *quatorze*; **German** *vierzehn*; **Spanish** *catorce*; **Italian** *quattordici*; **Latin** *quatturodecem*; **Greek** *tessares kaí deka*; **Papuan body counting,** left ear; **classical and medieval finger numerals**, left index finger touching last joint of thumb (sign for "ten") with ring and middle fingers folded at joints (sign for "four")

When a weekend or a week would be too short and a month too long, the English are fond of waiting for a fortnight, two weeks or fourteen days. But for another view of this issue, *see* **15**.

FIELD MARKS   The number of different ways that you can separate a hexagon into triangles that do not overlap by connecting vertices is fourteen (keeping track of each vertex). Here are four of them:

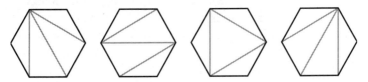

As you can see, the fourth is just like the first, but the common vertex is different. There are 6 like the first sample above, 6 like the second sample, and 2 like the third to make the total of 14. The problem of separating polygons into triangles is interesting in part because triangulation, or separating into triangles, is the method of measuring the area of polygons, even though we convert areas to squares when we report measurements. In high school geometry, however, you start with a square and show how the area of a rectangle (with natural-number sides) can be shown as a number of squares based on length and width. Then triangulation begins. First the rectangle is used to establish the area of the right triangle; then oblique triangles are broken into right triangles; and, to complete the triangulation, the areas of all further polygons are found by separation of the polygons into triangles.

The result stated for counting the line segments used in triangulating hexagons is an instance of a general rule that was studied by Euler. He found that the numbers form a sequence that begins 1, 2, 5, 14, and that has as the general term—starting with $n = 3$—

$$\frac{2 \times 6 \times 10 \times \cdots \times (4n - 10)}{(n - 1)!}$$

using the ! notation for factorial, or product of consecutive counting numbers (*see* **Factorial Family,** pp. 145–146). This is a complicated formula, and Euler had a terrible time discovering it. Surprisingly, if you allow a second 1 as the first term (a 1 that has no geometric counterpart, unfortunately), the sequence that begins 1, 1, 2, 5, 14, . . . and continues exactly as before has a much simpler iterative formation rule: if k is the member of the sequence preceding the nth member, then $k(4n - 6)/n$ is the value of the nth member.

This sequence reappeared in 1838 when the Belgian mathematician Eugène Charles Catalan [1814–1894] investigated the number of ways that a chain of entities can have parentheses inserted so that each operation is binary. While one entity makes no sense by itself, there is one way for two entities to be linked as a binary operation (here multiplication) in a given order, which requires no parentheses: ab. For a product of three entities there are the two familiar ways from the statement of the associative property: $a(bc) = (ab)c$. For four entities the number of ways is five: $(ab)(cd)$, $((ab)c)d$, $a(b(cd))$, $a(bc)d$, and $(a(bc))d$. For five entities, there are fourteen ways. As a result of this discovery, the sequence that begins 1, 2, 5, 14, . . . is known as the Catalan sequence and not the Euler sequence, which is just as well since the number of mathematical objects named after Euler is already too confusing.

After the figurate sequences (such as the triangular and square numbers) and the Fibonacci sequence, the Catalan sequence is perhaps the most commonly occurring in the mathematics of counting numbers. It shows up in several kinds of network problems as well as in other geometric contexts.

One of the many things named after Euler is his ϕ or indicator function, which is one of the most interesting functions of whole numbers. The ϕ function is defined over the natural numbers in such a way that $\phi(n)$ is the number of counting numbers less than n that are relatively prime to n. (Two numbers are relatively prime if their greatest common factor is 1.) In this case, the number 1 is treated as relatively prime to all other counting numbers, a rather special usage of the term. Also, $\phi(1)$ is considered to be 1. The next Euler indicator is $\phi(2) = 1$ since 2 is relatively prime to 1, the only counting number less than 2 in any case. But 3 counts both 1 and 2 as relatively prime, so $\phi(3) = 2$.

The values of ϕ seem to go up and down almost randomly—for example, $\phi(9) = 6$, $\phi(10) = 4$, and $\phi(11) = 10$—but Euler found various ways to derive ϕ that are basic to understanding many properties of prime numbers. One of the easier to spot is that if n is itself prime, then $\phi(n) = n - 1$. For example, the ten numbers less than or equal to 11 that are relatively prime to 11 are *all* the counting numbers less than or equal to 11. One of the more sophisticated results of using Euler's function is that if two counting numbers a and b are relatively prime, then $\phi(ab) = \phi(a)\phi(b)$, a result that is used in the "public key" method for encoding and decoding information.

Where does 14 enter the picture? There is no n such that $\phi(n) = 14$, and 14 is the smallest number that does not appear as a value of Euler's indicator function.

Also, 14 is a square pyramidal number (*see* **Square Pyramids**, p. 127)

SIMILAR SPECIES   The Catalan sequence continues 42, 132, 429, 1430, 4862, 16,796, . . .

PERSONALITY   Although 14 is factored as 2×7, its relationship to 1/7 is one of the most useful associations in quick calculations in mathematics. The reason is that 1/7 as a decimal begins 0.142 . . . and continues on, infinitely repeating (*see* discussion at **Genus *Rational*, 1/7**). The percentage version is $1/7 \times 14\frac{2}{7}\%$. Thus, a good approximation to multiplication by 14 is division by 7 followed by multiplication by 100. This may seem unhelpful, since operations with 7 are notoriously difficult, but for small numbers the relation is usable going in both directions. For example, multiplying 85 by 14 is easy if you think 85 is close to 84, and $84 = 7 \times 12$, so 85×14 is about $(84 \div 7) \times 10 = 1200$. (The actual product is 1190).

This approach is even more useful in division. For example, to find $500 \div 14$, multiply 500 by 7 instead of dividing by 14 (the answer will be 100 times what you need). So, $500 \div 14$ is about 35. Conversely, $500 \div 7$ is about 70 (think $14 \times 5 = 7 \times 10$).

ASSOCIATIONS   The British famously measure their personal weight in stone instead of pounds or kilograms. One stone is 14 pounds.

The best-known verse form in English (and Italian) is the sonnet, which is fourteen lines. Poetry is full of counting—of lines, syllables, and accents. The poet's preoccupation with counting (or "prosody") is satirized in this "Sonnet on Prosody" (which I wrote while a college English major):

> When I have fears my line will be too long,
> I measure off the iambs on my digits.
> Haltingly rhyme in sweet anapestical song,
> Shouting trochees until I have the fidgets,
> For a lark, turning out if I can a light dactyl or two;
> For each my fingers tap out five-foot forms.
> This guarantees a line with meter long and true,
> Since short ones may cause critics' storms.
> The poet must be master of his call
> And in each verse must work his will upon it;
> Besides the meter he must know its forms,
> The rules of rhyme—and always must recall
> That there are fourteen lines in every sonnet.

Square Pyramids

One of the easiest ways to stack balls is to make a square for each layer, so that a stack 10 balls high would have in each succeeding layer 100 balls, 81 balls, 64 balls, 49, 36, 25, 16, 9, 4, and 1 ball on top. The result would be a square pyramid, and so the numbers that are the sums of the squares taken in order are called the *square pyramidal numbers*. Starting with $1^2 = 1$, they are 1, $1^2 + 2^2 = 5$, $1^2 + 2^2 + 3^2 = 14$, $1^2 + 2^2 + 3^2 + 4^2 = 30$, $1^2 + 2^2 + 3^2 + 4^2 + 5^2 = 55$, and so forth. There is a formula for the nth square pyramidal number, which is

$$\frac{1}{6}n(n + 2)(2n + 1)$$

15 Fifteen XV 3 × 5

French *quinze;* **German** *fünfzehn;* **Spanish** *quince;* **Italian** *quindici;* **Latin** *quini deni;* **Greek** *pente kaí deka;* **Papuan body counting,** left shoulder; **classical and medieval finger numerals,** left index finger touching last joint of thumb (sign for "ten") with middle finger folded at joints (sign for "five")

One of the odd places to encounter 15 is in the French expression for two weeks, which is *quinze jours,* or fifteen days. This use is related to the ancient Germanic custom of counting nights instead of days, so two weeks—fourteen nights, or a fortnight in English—is fifteen days in French.

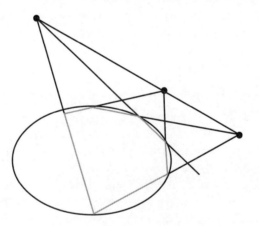

Pascal's Mystic Hexagram

FIELD MARKS A famous theorem in geometry is Pascal's mystic hexagram (*see also* discussion at **3**). Pascal showed that when a hexagon (other than a regular one) is inscribed in a conic section, the three points of intersection of the extensions of its opposite sides lie on a line. This theorem has been investigated by many mathematicians since Pascal, who himself reputedly found 400 corollaries to the theorem. In many cases the special points and lines identified come in sets of 15, including the fifteen so-called Salmon points and fifteen Plücker lines.

Other recognizable traits: 15 is a triangular number ($5^\triangle = 15$), part of the fourth smallest Pythagorean triple, 8, 15, 17, and the fourth Bell number (*see* discussion at **52**).

SIMILAR SPECIES   The multiples of 15 are useful to know: 30, 45, 60, 75, 90, 105, 120, and 135 (through 9×15).

PERSONALITY   Not only can 1/7 be used as a rough approximation in operations based on 14 (*see* discussion at **14**), but also for those with 15. For 15, it is slightly rougher since $1/7 = 0.142857 \ldots$ is closer to 14 than to 15. As people often tip waiters 15% or slightly higher, the tip can be calculated by division by 7, often an easier calculation than multiplication by 15. The quotient will be slightly lower than 15% of the bill.

Of course, multiplying by 15 can also be thought of as multiplication by $10 + 5$. Using this method, to find 15% of a bill of $32.27 you would begin by taking half of $32 (the nearest full dollar) and add that $16 to $32 to get $48. The tip should be more than $4.80, and $5.00 is a good round number ($4.80 is about 14.9% of the original $32.27 and $5.00 is about 15.5%).

The square of any number ending in 5 can be easily created by a method that may seem odd at first but becomes quite useful in practice. The square will always end in the digits 25 preceded by the product of the number represented by the digit(s) preceding the 5 with the number that is 1 more than that number. For 15, that means that the first part of the square is the product $1 \times (1 + 1)$, which is 2, so the square is 225. (The method even works for 5 alone, since the preceding digit is 0.) This method of squaring becomes more useful for greater numbers. For example, the first part of the square of 65 must be $6 \times (6 + 1) = 42$, so $65^2 = 4225$. The method works even for numbers greater than two digits that end in 5, such as 125, which begins with $12 \times 13 = 156$ and ends with 25, so $125^2 = 15,625$. Of course, the larger the number is, the harder this technique is to do in your head, but it can still make a paper-and-pencil calculation easier.

Why does this work? Notice that the actual square represented as a decimal number ending in 5 means $(10n + 5)^2$, which can be expanded by ordinary rules of algebra to $100n^2 + 100n + 25$, which factors as $100n(n + 1) + 25$. The first addend will always be n times $n + 1$ with two 0s annexed, while adding 25 will replace the digits 00 with 25.

ASSOCIATIONS   The Fifteen Puzzle is a well-known game in which the natural numbers from 1 through 15 are marked on tiles that can be moved about in a square frame that could hold sixteen such tiles. A tile cannot be picked up—it can move only by shifting into the blank space when that space is adjacent. The puzzle is solved when a scrambled numerical sequence is rearranged in counting-number order (from least to greatest). Scholars think that the Fifteen Puzzle was invented in the United States in 1878; by the early 1880s it was popular all over Western Europe as well as in its homeland.

If the puzzle has been set up by scrambling the existing tiles from a solution, without lifting any of them, it can always be solved. It was quickly realized, however, that lifting the tiles from the board and then scrambling them leads to configurations that defy any solution. As early as 1879 the theory behind this difficulty was worked out by several different mathematicians. It turns out that there are only two basic arrangements, and if two tiles are reversed in order from one arrangement, the structure changes to the other arrangement. In the early 1880s the American puzzle inventor Sam Loyd offered a prize to the first person who could solve a puzzle that he sold in which 14 and 15 had been interchanged from what otherwise would be counting-number order, knowing full well that his money was safe.

The earliest-known and smallest magic square has a sum through the rows, columns, and diagonals of 15 (*see* discussion at **9**).

> Fifteen men on the Dead Man's Chest
> Yo-ho-ho, and a bottle of rum!

As I was thinking about 15 and remembering my Robert Louis Stevenson, I could not decide whether the fifteen men were on the upper body of a deceased male or on a murdered pirate's treasure chest, so I looked it up. It was neither, but one of the Virgin Islands in the Caribbean, nicknamed Dead Man's Chest by English buccaneers.

16   Sixteen   XVI   2 × 2 × 2 × 2

French *seize*; **German** *sechzehn*; **Spanish** *dieciséis*; **Italian** *sedeici*; **Latin** *sedecim*; **Greek** *hekkaídeka*; **Papuan body counting,** left elbow; **classical and medieval finger numerals,** left index finger touching last joint of thumb (sign for "ten") with ring finger folded at joints (sign for "six")

The number 16 and multiples thereof are among the quantities most familiar to book designers and printers because most books at one time were printed on large sheets of paper that were folded three times and cut at the edges. The original large sheet has two sides, which in a book printed without folding would represent two pages. Each fold doubles the number of pages, so the first fold produces four pages, the second eight, and the third sixteen. As a result, even short books usually have sixteen, thirty-two, forty-eight, sixty-four, or ninety-six pages. (For some reason, eighty pages is not a popular choice.)

FIELD MARKS Since 16 is the fourth power of 2 ($2^4 = 16$) and also the second power of 4 ($4^2 = 16$), it is represented on both sides of the odd equation $n^m = m^n$ where n and m are both natural numbers. Indeed, if $n \neq m$ then 16 is the only number that can be on both sides of the equals sign.

The number 16 is the fifth of the *Euler numbers*, which give the number of ways that the beginning sequence of the counting numbers can be arranged so that they alternately rise and fall in size. For example, 12 or 35 are arrangements from the counting numbers 1, 2, 3, 4, and 5 that rise, while 21 and 42 are arrangements that fall. When you arrange the five numbers as 14352 they alternately rise and fall. There are sixteen ways to do this with five numbers, including 24351, 25341, 34251, and 15243, and so is why 16 is called the fifth Euler number.

Another name for the Euler numbers is *zigzag numbers*, which has its origin in a way of constructing these numbers from a sequence that zigzags across a number triangle; this name also suggests the alternating rise and fall of the number sequences in the definition. In this version the even-numbered Euler numbers are the zigs while the odd-numbered ones are the zags. Since 16 is the fifth Euler number, it is a zag.

Although it would seem unlikely that they could prove to be useful in any way, the Euler numbers reappear in infinite sequences that produce the trigonometric functions called the tangent and the secant. The connection is that the tangent of x is the infinite sum for which each term is the $(2n + 1)$th Euler number times x to the $(2n + 1)$th power over the factorial of $(2n + 1)$ (*see* **Factorial Family**, pp. 145–146). Notice that the Euler numbers in this definition are all zags. The secant of x has the same relation with the zigs.

The smallest cyclic number, with 6 digits, is 142,857. The next smallest cyclic number (*see* discussion at **6**) has 16 digits: 0,588,235,294,117,647. The 0 at the beginning is a feature of all cyclic numbers except the first, which is 142,857. The 0 is needed because, for example, $2 \times 588,235,294,117,647 = 1,176,470,588,235,294$. Although the 0 appears at another place, not at the beginning, in the multiples, it always hangs out at the place it should be in the cycle. For example, 5,882,352,941,176,470 is easily recognized as the basic number times 10.

Finally, 16 is part of the Pythagorean triple 16, 63, 65.

SIMILAR SPECIES Larger books are often printed on high-speed presses in 64s instead of 16s, starting with even larger sheets and giving them two folds (a high-speed web press prints from a continuous roll, but it still has to cut and fold to make pages). Thus 512, 576, and 704 pages often appear, or often $512 \pm 32 = 480$ or 544 pages. One kind of medium-sized press gives the paper a fold into thirds instead of halves, making it possible to print in forms of 48, resulting in less expense for 336 (seven 48s) or 432 (nine 48s) pages than twenty-one or twenty-seven forms of 16 pages.

Similar species to 16 considered as the fifth Euler number are the numbers for zigzag arrangements of 1, 2, 3, 4, 6, and 7 counting numbers: these are, respectively, 1, 1, 2, 5, 61, and 272.

PERSONALITY   The most important aspect of 16 for many situations is that it is the fourth power of 2, so it is easy to use by doubling or halving four times in a row. For example, to multiply 42 × 16 you could think 84, 168, 336, 672. Division works best when there are no fractions involved, that is, when the number being divided is a multiple of 16. For example, 848 ÷ 16 can be calculated as 424, 212, 106, 53, but 957 is a mess when you try the same approach.

A method that is often useful for approximate multiplication or division by 16 is to remember that 1/6 is the same as 0.16666 . . . or 16⅔%, so dividing by 6 and multiplying by 100 approximates multiplying by 16. For example, to multiply 42 × 16, think (42 ÷ 6) × 100 = 700—the exact answer is 672. Similarly, multiplying by 6 and dividing by 100 approximates dividing by 16. Whether or not one of these approaches will be helpful in a particular case depends on the numbers; it is not all that easy to multiply 957 by 6 in your head, for example, and you might prefer to divide by 16 using the ordinary algorithm.

ASSOCIATIONS   In 1998 the total number of knots that have sixteen crossings was calculated separately by a pair of mathematicians and by another working alone. They found the same result: there are exactly 1,701,936 such knots. The number 16 is the maximum number of crossings that have been computed so far, but no doubt that will be soon exceeded (*see* discussion at 3).

An avoirdupois pound, the common measure of weight in the United States, contains 16 ounces.

---

17 Seventeen XVII prime

French *dix-sept*; **German** *seibzehn*; **Spanish** *diecisiete*; **Italian** *diciassette*; **Latin** *septemdecim*; **Greek** *hepta kaí deka*; **Papuan body counting,** left wrist; **Classical and medieval finger numerals,** left index finger touching last joint of thumb (sign for "ten") with little finger folded at knuckle (sign for "seven")

Karl Friedrich Gauss was a skilled linguist as well as a mathematical genius. As a teenager he was not sure whether he should make his career in mathematics or in languages. When on March 30, 1796, he discovered a way to inscribe a regular seventeen-sided polygon into a circle—the first significant improvement in geometric constructions since the Greeks—he decided that he should stick with mathematics.

FIELD MARKS Fermat guessed that numbers formed by raising 2 to a power of 2 and adding 1 are all prime; these are the "Fermat primes." Probably Fermat observed that $2^1 + 1 = 3$, $2^2 + 1 = 5$, $2^4 + 1 = 17$, easy primes, and calculated $2^8 + 1 = 257$ and $2^{16} + 1 = 65,537$ and found that they were primes. Furthermore, he knew a related theorem that any number of the form $b^n + 1$ can be prime only if n is a power of 2. But Fermat was wrong: various different later mathematicians established that $2^{32} + 1$, $2^{64} + 1$, and so forth are composite. Today it is believed that the only Fermat primes are 3, 5, 17, 257, and 65,537.

Gauss's construction of a seventeen-sided polygon is related in an odd way to the Fermat primes. All the regular polygons with an odd number of sides that can be constructed with straightedge and compass are multiples of the Fermat primes or of the products of Fermat primes. As a result, the numbers of sides are all divisors of $3 \times 5 \times 17 \times 257 \times 65,537$, which is equal to $2^{32} - 1$

The number 17 is the only prime number that is the sum of four consecutive prime numbers. This is easy to establish when you realize that the sum of any four prime numbers will be odd unless one of the primes is 2.

Also, 17 is one of the primes that produce a cyclic number as the repeating period on division into any number that is not a multiple of 17 (*see* discussions at **6, 7,** and **Genus Rational, 1/9**).

Other recognizable traits: 17 is a prime twin with 19. It has been proposed that, since there can never be any prime triplets greater than 3, 5, 7, groups of primes such as 11, 13, 17, 19 be designated *prime quadruplets*. This means ignoring the multiple of 3 that will always insert itself in the middle (15, in this case).

Also, 17 is part of the Pythagorean triple 8, 15, 17.

Finally, 17 is the last consecutive natural number about which there is enough interesting to deserve a separate listing in this field guide. Although 18 is twice the square of 3 and half the square of twice 3, and also a magic number in chemistry (for a stable number of electrons), after that it gets to be rather dull.

SIMILAR SPECIES Some other numbers that produce the maximum-sized repeating period and lead to cyclic numbers

are 7, 19, 23, 29, 47, and 17,389. There are probably an infinite number of such primes, although this has not been proved.

The next prime quadruplet after 11, 13, 17, 19 is 101, 103, 107, 109.

PERSONALITY   Although 17 is a particularly awkward prime, as are the others that produce cyclic numbers, it is small enough that it is useful to memorize the smaller multiples, which are 34, 51, 68, 85, and 102. These are useful in many ways. In particular, knowing $34 = 17 \times 2$ is helpful because 34 itself often occurs as an obvious factor; recognizing that 51 is 17×3 keeps you from thinking that 51 is prime; and knowing $85 = 17 \times 5$ is of considerable help in unraveling multiples of 5.

ASSOCIATIONS   The number 17 appears in a now rejected claim by the distinguished physicist Arthur Eddington [English, 1882–1944] that the *exact* number of protons in the universe is 17×2^{259}.

---

19   Nineteen   XIX   prime

French *dix-neuf*; **German** *neunzehn*; **Spanish** *diecinueve*; **Italian** *diecinove*; **Latin** *undeviginti*; **Greek** *ennea kaí deka*; **Papuan body counting,** left index finger; **classical and medieval finger numerals,** left index finger touching last joint of thumb (sign for "ten") with little, ring, and middle fingers folded at knuckles (sign for "nine")

In 432 B.C. the Greek astronomer Meton observed that the number of days in 235 cycles of the phases of the moon is almost exactly equal to nineteen times the number of days in the year. Thus, the first day of the seasons can be kept regular by using a lunar calendar based on repeating every nineteen years. To make each year of lunar months about the same as the true year based on the revolution of Earth about the Sun, you need to have about twelve or thirteen lunar months each year, so the Metonic calendar contains twelve years with twelve lunar months and seven years with thirteen months to produce the required 235 lunar months every nineteen years. This was the main calendar of Western civilization until Julius Caesar's calendar reform of 46 B.C., and it is still the basis of the Jewish calendar. A similar calendar was developed independently by Babylonian astronomers about the same time as the Metonic.

FIELD MARKS   The number formed by stringing nineteen 1s together, 1,111,111,111,111,111,111, is the first rep-unit after 11 that is prime (*see* discussion at **11**).

SIMILAR SPECIES   A century after Meton, the Greek astronomer Callipus improved the Metonic cycle of nineteen years, which drifted by a day after four cycles, by developing a calendar based on repeating the lunar cycle every seventy-six years. Callipus' system was preferred to Meton's by later Greek astronomers, although Hipparchus showed that a system based on 304 years and 3760 lunar months is even more accurate.

PERSONALITY   Although 19 is a prime, its proximity to 20 makes multiplication with 19 as a factor easy. Also, it is useful to know the products of 19 with 3 and 7, since they may easily be mistaken for primes: $19 \times 3 = 57$ and $19 \times 7 = 133$.

20 Twenty XX $2 \times 2 \times 5$

French *vingt;* **German** *zwölft;* **Spanish** *viente;* **Italian** *venti;* **Latin** *viginti;* **Greek** *eíkosi;* **Papuan body counting,** left middle finger; **classical and medieval finger numerals,** left thumb touching base of index finger

The original score was a tally mark, named from the Old Norse *skor,* which led to the use of the word "score" for 20 as well as to its meaning as the tally in a contest. Lincoln famously calculated 87 years as "four score and seven." If you score a piece of glass or wood so that it will break in a straight line, it is the same word again—you are making a mark as you would on a tally stick. Until the late eighteenth century, the main financial records of the British Parliament were kept on wooden tally sticks. After paper records had been in use for some decades, it was thought safe to destroy the old wood sticks. They were burned in 1834, in a fire that also consumed the Houses of Parliament.

FIELD MARKS The mystic hexagram of Pascal (*see* discussions at **3** and **15**) has many sets of 20 points or lines resulting from it that have been investigated by various mathematicians. These include the 20 Steiner points that lie on the 15 Plücker lines and which each have 3 Kirman points associated with them that lie on the 20 Cayley lines which in turn pass through the 15 Salmon points (*see also* **3** and **60**).

Also, 20 is one of the legs of the 20, 21, 29 Pythagorean triple as well as of the triple 20, 99, 101.

SIMILAR SPECIES 20 is to 100 as 1 is to 5, a relationship that is frequently useful.

PERSONALITY   A tip of 20% is often more easily calculated by dividing the bill by 5 instead of doubling and moving the decimal point. For example, a bill of $35 would result in a tip of $7 since $35 \div 5 = 7$. The alternative computation is to double 35 to get 70, then drop the final 0 to get 7.

ASSOCIATIONS   In geometry, 20 is the number of faces of an icosahedron and therefore the number of vertices of a dodecahedron (*see* illustration at **5**).

The number 20 is one of the magic numbers for radioactive decay—meaning that an atom with twenty nucleons (the main isotope of neon) is particularly stable. Neon is doubly magic, since it also has ten electrons, a magic number for chemical stability. Hot sodium atoms also tend to form clusters of twenty atoms, making this a magic number for clusters as well.

The old English pound equaled 20 shillings, but nowadays it is divided into 100 pence. At one time a gold version of the pound called the guinea, worth 21 shillings, was in circulation. The odd amount was standardized in 1717 at 21 so that an auctioneer could accept bids in pounds, be paid in guineas, and pocket the difference as his fee. A sporting way to raise a bet or an offer in England was to say "guineas" to increase stakes slightly. The guinea was minted in gold from 1663 to about 1813, so it is most familiar today to readers of eighteenth- and nineteenth-century novels.

---

22   Twenty-Two   XXII   2 × 11

FIELD MARKS   The number 22 is the first of the two-digit Smith numbers and the lowest after 4. A *Smith number* is one whose digits add to the same number as the sum of the digits in all its prime factors. Note that the sum of the digits in both 4 and 22 is 4, while 4 factored into primes is 2 × 2 (sum of digits is 4) and 22 is factored as 2 × 11 (sum of digits is 4).

SIMILAR SPECIES   The next few Smith numbers are 27, 58, 85, 94, and 121. Some Smith numbers that are famous for other reasons are 666, the "Number of the Beast" in the Bible, and 1776, the birth year of the United States (the sum of the digits of 1776 is 21; its factors are 2 × 2 × 2 × 2 × 3 × 37 (and the sum of those digits is 2 + 2 + 2 + 2 + 3 + 3 + 7 = 21).

PERSONALITY   Since 22 is twice 11 as well as close to both 20 and 25, computation or estimation using it is a snap. Even using the standard algorithm for multiplication is easy, especially if you double the entire other factor instead of taking it one digit at a time. For example, to multiply 22 × 247 you can think "the double of 247 is 494" (this can also be found by recognizing that 247 is 3 less than 250 so twice 247 is 6 less than 500). Then write

$$\begin{array}{r} 494 \\ \underline{4940} \end{array}$$

and add to get the product of 5434.

ASSOCIATIONS   The English surveyor's chain is set at a standard of 66 feet or 22 yards; it has 100 links, hence each link is 7.92 inches long. These peculiar numbers appear to be related to the rod, which at 5½ yards is exactly one fourth of 22 yards (*see also* discussion at **40**.) Ten chains, or 220 yards, make a furlong, a distance most familiar today as a length for various kinds of races—the "220" and its related "440" for track and multiples of furlongs for horse races.

23   Twenty-Three   XXIII   prime

FIELD MARKS   The poet and mathematician Sun-tse [Chinese, *fl.* fourth or fifth century A.D.) asked for the smallest natural number that when divided by 3, 5, and 7 will produce the respective remainders 2, 3, and 2. The number is 23—that is, $23 \div 3 = 7$ r 2; $23 \div 5 = 4$ r 3; and $23 \div 7 = 3$ r 2. Sun-tse supports his answer with a poem about the general method. The general theorem, called the Chinese Remainder Theorem, is of great importance in number theory and in the foundations of mathematics. The Chinese Remainder Theorem states an algorithm that provides a number that gives a specified set of remainders when divided by some given numbers that are relatively prime in pairs.

The rep-unit formed by stringing together 23 1s is a prime (*see* discussion at **11**).

SIMILAR SPECIES   Although 23 is a larger prime than 17 or 19, the small size of both its digits makes the smaller multiples of 23 easier to spot than small multiples of 17 or 19. It is easy to recognize 46, 69, and even 92 as multiples of 23.

PERSONALITY   In most cases of multiplication by 23 it is easiest to use a version of the standard algorithm that begins by doubling the other factor, which can be done mentally. Write that number on a slip of paper, put the original number under it, then put the doubled form off-set one unit to the left and add the three numbers. For example, to find 23×638, think 638 doubled is 1276 and write

142   Kingdom Number

$$\begin{array}{r} 1276 \\ 638 \\ 1276 \\ \hline 14674 \end{array}$$

So the answer is 14,674. Notice that this is essentially the same as using 22 as a friendly number and treating the problem as $(22 + 1) \times 638$.

ASSOCIATIONS In 1900 mathematician David Hilbert [German, 1862–1943] gave a famous speech to the International Congress of Mathematicians in Paris in which he set forth twenty-three unsolved problems of mathematics, problems that he hoped would be a main focus of the subject in the twentieth century. Since Hilbert's speech, progress in many parts of mathematics, especially in understanding its foundations, has been measured by how far we have progressed toward solving these problems or in proving them insoluble. For the most part, these problems require mathematical ideas that are beyond the scope of this book. One problem, however, concerns number families. He asked whether $2^{\sqrt{2}}$ is transcendental or algebraic (*see* **Transcendental Family,** pp. 290–292)—that is, whether or not it can be stated as a solution to an algebraic equation. Hilbert thought this problem would not be solved during his lifetime, but a student who heard Hilbert's speech proved only a decade later that $2^{\sqrt{2}}$ is transcendental. A student of Hilbert's was the first to prove any of the problems, solving the third one within a couple of years of the speech. Several problems were "solved" by proving that they could not be solved by mathematical means. Some problems, however, remain unresolved to this day, at the close of the twentieth century.

The number 23 is also one of the two numbers that Freud's friend Wilhelm Fliess believed to hold the secrets of life (the other was 28). Fliess thought that 23 was the natural rhythm of cells in males, while 28 was the natural rhythm of females (not the menstrual cycle, he said; that was an effect, not a cause). This idea persists to some degree under the name of biorhythms. There is no evidence in support of 23 as a rhythm of life.

24   Twenty-Four   XXIV $2 \times 2 \times 2 \times 3$

FIELD MARKS   The number 24 is a factorial ($4! = 24$—*see* **Factorial Family,** pp. 145–146), and it is also the smallest factorial that becomes a perfect square by the addition of 1, since $4! + 1 = 5^2$.

Also, 24 is a member of the Pythagorean triple 7, 24, 25.

SIMILAR SPECIES   There are two other known factorials that become perfect squares upon adding 1, both quite small. In fact, one of them is the next after 4! since $5! = 120$ and $121 = 11^2$. The other known example is $7! = 5040$. The number 5041 is the square of 71.

The number 5040 is also called Plato's number, because Plato suggested that this was the perfect size for a community, mainly because 5040 has 58 divisors (not including 1 or 5040 itself), which is rather large (but *see* discussion at **6**.)

PERSONALITY   The product of any four consecutive integers is a multiple of 24. This occurs because one integer out of any four in a row must be a multiple of 4, while another must be a multiple of 3. This combination gives a factor of 12, but there are two even numbers in any four consecutive integers, so in addition to the factor of 4 there is a factor of

2. Put them all together and you get 24. For example, 13 ×
14 × 15 × 16 is 13 × (2 × 7) × (3 × 5) × (4 × 4) = 13 ×
7 × 5 × 4 × (2 × 3 × 4) = 13 × 7 × 5 × 4 × (24).

ASSOCIATIONS A pennyweight is 24 troy grains, which is
1/20 of a troy ounce. As its name suggests, it was once the
weight of a penny, when English pennies were silver.

Factorial Family

Numbers that are the products of consecutive counting numbers, starting with 1, are called *factorials*. It is easy to picture 2
factorial as 1 × 2 = 2 or 3 factorial as 1 × 2 × 3 = 6, but
what of 1? Many problems involving counting arrangements
or combinations of things can be solved with formulas involving factorials, and these formulas give correct answers if 1 factorial and even 0 factorial are each assigned the value 1.

It is awkward to write "factorial" every time, and furthermore the mathematician wants some simple way to
insert factorials in equations. The first notation mathematicians tried was a sort of half box, writing 3⌋ for 3 factorial.
Not only did this fail to suggest the nature of the factorial,
but it was difficult to print in complicated formulas involving several factorials and fractions, so in 1808 Christian
Kramp [German, 1760–1826] introduced a new sign—the
only use of the exclamation point in mathematics. As a
result of Kramp's idea, factorials have come to seem exciting.
0! = 1, 1! = 1, 2! = 2, 3! = 6, and 4! = 24. The exclamation point happily suggests how fast the factorial function
rises, much faster than any power or indeed than any other
commonly used function.

Mathematicians still *say* factorial, of course, reading the symbol 4! aloud as "four factorial." I have often thought it would be better if they used the verbalization of ! that proofreaders use when they read an important document aloud to a copyholder to make sure there are no errors in transcription. "Dammit!" would be read as *quote cap D dammit d-a-m-m-i-t bang close quote*. The exclamation point becomes a "bang"! That would suit mathematics, in my opinion. For instance, the number of combinations of n things taken r at a time has the formula n!/r!(n − r)!, which a proofreader would pronounce as "n bang over r bang times the quantity n minus r bang." This is not, however, accepted in mathematical circles.

26   Twenty-Six   XXVI   2 × 13

FIELD MARKS   The number of primes in a given interval declines irregularly, but seems to go down as one gets to higher numbers, which is what one might expect. For example, the first twenty-six prime numbers, from 2 to 101, occupy an interval of exactly 100 numbers. The next interval of 100 numbers, from 102 to 203, contains only twenty primes. Thereafter, the numbers per 100 tend decline, but not steadily: there are sixteen each in the 200s and 300s, an increase to seventeen in the 400s, down to fourteen in the 500s, up to sixteen in the 600s, down to fourteen again in the 700s, up to fifteen in the 800s, and back to fourteen in the 900s. At higher numbers, however, the reduction is more noticeable. Between 100,001 and 100,100 there are only six primes. As a result of this decline, the twenty-six primes from 2 to 101 are the largest number of primes in any interval of one hundred consecutive counting numbers.

146   Kingdom Number

SIMILAR SPECIES   There are 168 primes in the interval from 1 to 1000.

PERSONALITY   In exact multiplication or division calculations, 26 is seen most often in terms of 13 and 52, which are half 26 and twice 26. It can also be helpful to remember that 104 is 4 × 26.

For approximate calculations, the proximity of 26 to 25 is helpful, especially since 25 is a fourth of 100, so an exact calculation with 25 can be found by multiplying by 100 (annexing two 0s) and dividing by 4.

For multiplying large counting numbers by 26, think of 26 as 25 + 1. To find, say, 384 × 26, convert the problem to 384 × (25 + 1) = (384 × 25) + 384. The mental operations (or written ones if you choose) are 384 × 25 = 38,400 ÷ 4 = 9600 and 9600 + 384 = 9984, which is the answer.

ASSOCIATIONS   If we recognize 26 at all, it is usually as half the number of weeks in a year and a good substitute for half a year (twenty-six weeks is 182 days, while half a leap year is 183 days). If you compare six actual months to twenty-six weeks in a non-leap year, however, there are only two periods of six months (October through March and December through May) that are exactly twenty-six weeks long. A leap year does better, because there are twice as many six-month periods that equal exactly twenty-six weeks (you can find these if you play around a bit).

28   Twenty-Eight   XXVIII   2 × 2 × 7

FIELD MARKS   The number 28 is the second perfect number (*see* discussion at **6**), since its factors less than itself are 1, 2, 4, 7, and 14, and 1 + 2 + 4 + 7 + 14 = 28.

Other recognizable traits: 28 is a triangular number ($7^\triangle = 28$) and a member of the Pythagorean triple 28, 45, 53.

SIMILAR SPECIES   In addition to 6 and 28, the numbers 496 and 8128 are small perfect numbers. Perfect numbers have an odd relationship to Mersenne primes (*see* **Mersenne Family,** pp. 85–87), since if M is a prime number of the form $2^p - 1$ where p is also prime—this is the definition of a Mersenne prime—then M(M + 1)/2 is a perfect number. The formula produces 6 for $p = 2$; 28 for $p = 3$; 496 for $p = 5$; and so forth.

If you remember that the formula for a triangular number is $n(n + 1)/2$, you will note that perfect numbers formed this way are also all triangular numbers.

ASSOCIATIONS   The number 28 figures in one of the more amazing coincidences in the history of astronomy. In 1776 the astronomer Johann Elert Bode [German, 1747–1826] published the discovery made four years earlier by a less famous German, Johann Daniel Titius [1729–1796], that the distance of each of the planets from the Sun is determined by a simple number sequence, a sequence now called Bode's law. The sequence is formed with doubling as a first step, beginning 0, 3, 6, 12, 24, 48, 96, 192, 384, and then adding 4 to each number, producing 4, 7, 10, 16, 28, 52, 100, 196, 388. What Titius observed and Bode made famous is that if the distance from Earth to the Sun is taken to be 10 units (third number in sequence, for the third planet), then 4 and 7 correctly describe the relative distances of Mercury and Venus from the Sun, while 16 is the correct number for the distance of Mars from the Sun. Jupiter, however, the next known planet, was not at 28 but at 52. The sequence was more firmly grounded with Saturn at 100.

Bode's law might never have become famous, but in 1781, just five years after it was published, William Herschel found Uranus, which fit the rule, being within a reasonable margin of error of 196. But what about that 28? Surely if the law predicted it, there should also be a planet there—although no one had a clue as to why Bode's law was successful.

Astronomers scoured the skies seeking the missing planet. On New Year's Day, 1801, the first day of the nineteenth century, the Italian astronomer Giuseppe Piazzi found a previously unknown, but very small, "planet." Karl Friedrich Gauss devoted many hours to calculating its orbit, in the process inventing the methods needed to obtain an orbit from a limited number of observations. From Gauss's orbit, the distance from the Sun is an easy consequence of the laws of gravity. It was 28 on Bode's sequence. But the "planet" was Ceres, now known as the largest asteroid. Soon, however, astronomers found other asteroids, all about 28 units from the Sun. They came to believe (incorrectly, as it was later shown) that a large planet had occupied the orbit at 28 that was decreed by Bode's law.

Then people began to expect a planet at 388, the next distance invoked by the law. But astronomers who used gravitational theory instead of number mysticism correctly predicted that the next planet, now known as Neptune, would be elsewhere. Instead of being at 388 units from the Sun, Neptune is at 300 units, while Pluto is even farther away from the Bode's number. Apparently the remarkable fit of Bode's law to Uranus at 196 and Ceres at 28 is a coincidence and nothing else.

The number 28 seems to garner strange, unprovable, and incorrect theories. It is also one of the two numbers (the other is 23) that Freud's friend Wilhelm Fliess believed to be the basis of life (see discussion at 23).

The number 28 is one of the magic numbers of nucleons (protons and neutrons) that discourage radioactive decay. Silicon is the atom whose isotope, with 28 nucleons, benefits from this bit of magic.

36   Thirty-Six   XXXVI $2 \times 2 \times 3 \times 3$

FIELD MARKS   The number 36 is the first nontrivial number that is both a triangular number and a square number, since $8^{\triangle} = 36$ and $6^2 = 36$.

Also, 36 is a member of the Pythagorean triple 36, 77, 85.

SIMILAR SPECIES   Other numbers that are also triangles and squares (after the trivial example of 1) begin with $1225 = 35^2 = 49^{\triangle}$; $41,616 = 204^2 = 288^{\triangle}$; $1,413,721 = 1189^2 = 1681^{\triangle}$; and continue infinitely.

ASSOCIATIONS   Euler became interested in 1782 in a problem now known as the Thirty-Six Officers: if there are thirty-six officers, one from each of six different ranks and one of each from six different regiments, can they march in a square formation so that each row and column contains exactly one officer from each rank and from each regiment? This is harder than it looks, and Euler did not solve it.

Such arrangements into rows and columns by category are called Latin squares (sometimes Euler squares). Arranging the same group by two different criteria is a typical problem in Latin squares, with the arrangements considered to be "mutually orthogonal," which is to say, in more familiar language, perpendicular to each other. The rather unlikely reason cited in reference books for the odd name is that in Euler's time such problems were shown using Latin letters such as a, b, c, . . . instead of Arabic numerals such as 1, 2, 3, . . .

150   Kingdom Number

As to the solution to the Thirty-Six Officers Problem, it comes from what seems like another mathematical world entirely. Projective geometry, complete with a "point or line at infinity," was originally invented as a method by which three-dimensional scenes could be shown correctly in a painting on a two-dimensional surface. This invention was first generalized by mathematicians for transformations of geometric figures and later redeveloped in terms of finite sets of points, called projective planes of order n. The existence of mutually orthogonal pairs of Latin squares was shown to be equivalent to the existence of certain finite projective planes. In 1900 the mathematician G. Tarry showed that there is no projective plane of order 6, which in turn implies that the Thirty-Six Officers Problem cannot be solved.

The number 36 is a magic number for chemical stability. Electron shells begin with two electrons nearest the nucleus. When these are filled, the next shell has eight electrons, as does the third shell. But the fourth shell matches the eighteen electrons in the first three shells with another eighteen of its own. The result is a magic number of $18 + 18 = 36$. The element with this atomic number is krypton, which is completely stable (and not to be confused with Superman's "kryptonite," which is magic of quite another kind).

There are various British measures that are seldom if ever used anymore. A chaldron of coal was 36 bushels, for example, an amount that weighed somewhat more than one long ton (*see* discussion at 12). A barrel of ale contains 36 gallons (but a wine barrel has only 31½ gallons). If one is measuring straw, a truss is a bundle of 36 pounds, and 36 truss add up to one load of straw. But there is a lot more hay—56 pounds or 60 pounds, depending on whether it has been cured or is new—in a truss of hay.

40   Forty   XL $2 \times 2 \times 2 \times 5$

FIELD MARKS   Euler discovered a remarkable formula, $n^2 - n + 41$, that gives prime numbers when n is any of the first forty natural numbers. I have read that Euler believed that the formula would always produce primes, but this seems impossible since it is easy to see that if $n = 41$ the result cannot be prime. But the first forty counting numbers do produce primes. This is a result of a property of the Heegner numbers (*see* discussions at **9** and at **Genus *Complex, i***).

Also, 40 is a member of the Pythagorean triple 9, 40, 41.

SIMILAR SPECIES   The 40 primes that derive from $n^2 - n + 41$ are 41, 43, 47, 53, 61, 71, 83, 97, 113, 131, 151, 173, 197, 223, 251, 281, 313, 347, 383, 421, 461, 503, 547, 593, 641, 691, 743, 797, 853, 911, 971, 1033, 1097, 1163, 1231, 1301, 1373, 1447, 1523, and 1601.

The formula $n^2 - n + p$ where p is some unspecified prime will generate a list of primes up to $p - 2$ for a few numbers other than 41. The numbers p for which this is true are called the Lucky Numbers of Euler. In addition to 41, the Lucky Numbers are 2, 3, 5, 11, and 17. The proof that these are the only ones is another consequence of the unusual properties of the mysterious Heegner numbers.

PERSONALITY   It is often useful to remember that 40 is 2/5 of 100, especially in percent problems; also it is helpful to note that 40 is 1/5 of 200.

ASSOCIATIONS   The number 40 appears in many odd common expressions and folk slang. Why does forty winks mean a short nap? What prompted the goal of 40 acres and a mule for freed slaves after the American Civil War? An English centipede is familiarly called a forty-legs, which is actually more accurate than the hundred feet of "centi-ped." An old slang expression for cheap whisky is 40-rod, since it was so powerful that it could kill from 40 rods away. There is an herb called the forty-knot, while the Tasmanian diamond bird is the forty-spot.

Of these, "40 acres and a mule" has a relatively simple and known explanation. Land areas in British and U.S. customary measure were measured in rods, a length of 16½ feet or 5½ yards. (In the United States we have settled on the name "rod" for this surveyor's length, but the British are apt to call it a "pole" or even a "perch.") Surveyors seem to make no distinction between a rod of length and a square rod of area, but I will. The whole measurement system for land is based on numbers related to 40. This is not an accident. If you make a larger square ("big square") from small ones, you must use at least four small ones. Then if you make a still larger square ("great square") from that big square the great square will have sixteen of the original small squares. Now, since we use a decimal system, make everything ten times as large. From this kind of surveying, the acre of 160 square rods arises. If you start with the acre and repeat the process of making larger squares one more time to obtain a square of 640 acres, you have a square mile. A mile square (that is, an actual square 1 mile on a side as opposed to any irregular region with an area of 1 square mile) is a section of land, the way that land is laid out on tax maps. A comfortable size for a prosperous family farm in the nineteenth century was a quarter section, or 160 acres, but it was once possible to eke out a living on a quarter of that amount of land, which was 40 acres. The mule was simply the most efficient farm animal if you have only one animal to pull your plow.

The Bible is rife with references to 40, ranging from Noah's big rain to the time Jesus spent in the desert being tempted by Satan. The season of Lent is the forty days before Easter. The original quarantine against the Black Death was *quaranta giorni*, or forty days. This Judeo-Christian 40 derives from a Semitic tendency toward using 40 as a general round number as 100 is used in the West. The example of Ali Baba and the forty thieves is from the same tradition.

The number 40 is also one of the mysterious magic numbers for clusters of atoms. Clusters of forty atoms have more stability and are therefore more likely to form than, say, those with thirty-nine or forty-two atoms.

48   Forty-Eight   XLVIII $2 \times 2 \times 2 \times 2 \times 3$

FIELD MARKS   In 1932 it was established that for 48 or any whole number greater than 48, there is a prime number that is between that number and that number plus an eighth of the number. For example, for 48, the higher limit is $48 + 6 = 54$ and the prime in that interval is 53.

Other recognizable traits: 48 is a member of the Pythagorean triple 48, 55, 73.

PERSONALITY   Because 48 is 3×2^4 it has many appearances (*see*, for examples, the discussions at **16** of folding paper and at **6** of numbers with many small divisors.) The divisors of 48 including the unit are 1, 2, 3, 4, 6, 8, 12, 16, 24, putting it among the more useful numbers for many purposes. For multiplication, however, its nearness to 50 is the more useful connection.

ASSOCIATIONS Although there are 880 magic squares of order 4 (*see* discussion at **9**), the remarkable ones are forty-eight magic squares that are called most-perfect pandiagonals. All magic squares have the property that their rows, columns, and diagonals add to the same number, sometimes called the magic constant or the magic number for the square. In the square below, the magic constant is 30.

0	13	6	11
7	10	1	12
9	4	15	2
14	3	8	5

A Most-Perfect Pandiagonal Magic Square

This magic square is also a pandiagonal, which means that its "short" diagonals such as 9, 10, 6 can be extended by adding the far corner, 5, to obtain the same magic constant. Similarly, 3 + 15 + 12 + 0 = 30; 8 + 4 + 7 + 11 = 30; and 2 + 1 + 13 + 14 = 30. Look also at the even shorter diagonals such as the combination 7, 13 and 8, 2 or the combination 3, 9 and 12, 6. Same sum.

But this is also one of the forty-eight most-perfect pandiagonals of order 4 (actually all the pandiagonals of order 4 are "most-perfect"; higher orders can have rather less perfect pandiagonals). Notice that if you were to use this magic square to tile a floor, keeping the numbers in the same orientation, you would have the floor covered in a continuous pattern using the first sixteen whole numbers. In that pattern, any contiguous four numbers in a straight line, horizontal, vertical, or diagonal would sum to 30. Furthermore, if you choose any four numbers forming a small square, their sum will also be 30. Even if you add four numbers along a diagonal so that each pair of numbers is separated by a third number, the sum will be 30. Most perfect indeed!

And there are forty-seven more of these where that came from.

Amazingly, higher orders can produce many more of such wonders, although the order must always be a multiple of 4. There are 368,640 most perfect pandiagonal magic squares of order 8 and 2,229,530,000 of order 12, with the number rising rapidly above that.

---

50 Fifty L $2 \times 5 \times 5$

FIELD MARKS   As with 5, the number 50 occurs implicitly in all measurement situations involving a measurement to the nearest 100. In such measurement, the correct interpretation of the observation that Texas is 800 miles across from east to west is that the distance is between $800 - 50 = 750$ miles and $800 + 50 = 850$ miles.

SIMILAR SPECIES   The same measurement connection occurs with multiples of 5 in general. For example, to the nearest 10 miles Texas is 760 miles across, or between 755 and 765 miles.

PERSONALITY   As half of 100, 50 is especially easy to handle for multiplication. You should also remember that division by 50 can be accomplished by doubling and keeping track of the decimal point. For example, to divide 836 by 50, double to get 1672 and then divide by 100 (move the decimal point two places to the left) to get the answer of 16.72.

ASSOCIATIONS Pentecost, when the Holy Spirit descended on the apostles, took place fifty days after Easter; the feast is now always celebrated on a Sunday seven weeks (forty-nine days) after Easter. In England Pentecost is a popular date for baptism, at which the newly baptized are dressed in white—hence, another name for the day is White Sunday, or Whitsunday.

The number 50 is one of the magic numbers of nuclear stability. Two elements with isotopes of 50 nucleons (protons and neutrons) are titanium and vanadium—but while titanium-50 is stable, as are several lighter isotopes of titanium, vanadium-50 is less stable than, and therefore much less abundant than, vanadium-51. Go figure.

The magic qualities of the nuclear numbers become less obvious as the size of the numbers increases. The theory explaining why these particular numbers should be magic is not entirely satisfactory in any case.

---

52   Fifty-Two   LII $2 \times 2 \times 13$

FIELD MARKS   The number 52 is the fifth Bell number, or the number of ways that five different objects can be arranged in distinct sets. The numbers are named after Eric Temple Bell, who is primarily known as a historian of mathematics, who investigated these numbers.

SIMILAR SPECIES Other small Bell numbers are 1, 2, 5, 15, and 203. For example, the sixth Bell number is 203, meaning that six distinct objects can be arranged in 203 different ways (*see* discussion at **11**).

PERSONALITY The relation of 52 weeks to 1 year is often a factor in business calculations. Since the relation of 52 to the year and the week derives ultimately from astronomy, there is no particular reason why we should be blessed with a number of weeks in the year that is divisible by 4, but we are. If Earth were slightly farther from the Sun, for example, the year might have 372 days and there would be 53 weeks in the year, a prime. However, a year a week shorter than our existing one would have 51 weeks, and accountants would likely separate the year into thirds of 17 weeks instead of quarters of 13 weeks.

As it is, careful accountants do something much more complicated than simply figuring 13 weeks to a quarter. The quarters are treated in terms of 3 months at a time. Only the second quarter of April–May–June is a perfect quarter of 91 days ($13 \times 7 = 91$), while the first quarter is usually a day short, and the last two are each a day long. So some accountants make adjustments as indicated by the number of days.

In computation, the relation of 4 and 13 to 52 is useful whenever these particular numbers are involved, although not particularly helpful as an aid to other calculations.

ASSOCIATIONS The reason that there are fifty-two weeks in a year is that $364 \div 7 = 52$. The 7 for days in a week is a whole-number approximation of the time between phases of the Moon; 364 is the closest multiple of 7 to the year of 365 days.

A deck of playing cards has fifty-two cards, not counting any jokers. The 52 comes from four suits of thirteen cards each. The number of cards in the deck was not always 52, however. Early Tarot cards had five groups of ten cards each; the Venetian Tarot had four suites of fourteen cards not unlike the suits of a modern deck of playing cards, except that instead of the jack there were two similar figures called the cavalier and the valet. Used for casting fortunes, the Venetian Tarot also included twenty-two picture cards that represented such figures as the Sun, Death, Justice—and the Fool, our modern joker. While the total of seventy-eight cards seems excessive, the Florentines added another nineteen fortune-telling cards to bring the total to ninety-seven. The French developed the precursors of the modern suits of spades, hearts, diamonds, and clubs. Keeping the four French suits and dropping all the fortune-telling cards (except, sometimes, the joker) gives the modern deck of 52 (or 53) cards.

65   Sixty-Five   LXV 5×13

FIELD MARKS   Although 65 is instantly recognizable as a multiple of 5, it is somehow surprising to find that it is 13×5. It may help to relate it to 52, the number of weeks in a year, so five "quarters" is $52 + 13$ or 65.

Also, 65 is part of three Pythagorean triples, 16, 63, 65 and 33, 56, 65 and 65, 72, 97. Therefore, 65 is the smallest hypotenuse of two different right triangles that have sides that are natural numbers, since $65^2 = 63^2 + 16^2$ and also $65^2 = 56^2 + 33^2$.

SIMILAR SPECIES   The multiples of 5 with other medium-sized primes, including 85, 95, 115, 145, 155, and

185, have many of the characteristics of 65. They are not as easy to handle arithmetically as multiples of 5 that have a composite factor.

PERSONALITY In addition to having the virtue of a factor of 5 and the flaw of a factor of 13, 65 is comfortably nestled between 64, which is the sixth power of 2, and 66, which contains an 11. So you have many choices when dealing with computations based on 65. Multiplication by any even number can use the 2 from the evenness and the 5 from 65 to become a 10, so 36×65 is the same as $18 \times 10 \times 13$, which may not seem to be an improvement until you mentally rework it as 260×9, which is (of course) $2600 - 260 = 2340$.

ASSOCIATIONS The association most Americans have with 65 is the age of "senior citizenship." It once was the age of retirement, but now people retire anywhere from their fifties on and many prefer not to retire at all. Oddly, as the age for the start of full social security payments in our society is on the increase, the age at which a person can get senior-citizen discounts has dropped in many stores, and is often sixty-two or even sixty. The AARP, America's union of retirees, starts sending you mail when you turn fifty, retired or not.

---

100   One Hundred   C $2 \times 2 \times 5 \times 5$

French *cent*; **German** *hundert*; **Spanish** *ciento*; **Italian** *cento*; **Latin** *centum*; **Greek** *hekatón*

FIELD MARKS Since 100 is a power of 10 (10^2), it is an important part of the decimal system of numeration, but as a number it seems to have no particular distinction. If we used a numeration system based on 12, for example, the number we now write as 100 would be written as 84, meaning (8 × the base) + 4 (in decimal notation, 96 + 4) and no longer look like anything special. But 100 is not a prime, not a figurate number, not a member of any especially interesting sequence (although it is a perfect square), not a factorial, and so on and on.

SIMILAR SPECIES Unlike 100, the number 10 is interesting no matter what system of numeration is used (*see* discussion at **10**), but higher powers of 10 are as dull as 100.

PERSONALITY Despite its ordinariness as a number, its special role in our numeration system makes 100 a vital aid in computation. Any number that is near to 100 can be easily computed by converting the number to 100 plus-or-minus the difference. Thus, in computation a number such as 97 can be replaced by $100 - 3$, and 97×57, for example, is the same as $(100 - 3) \times 57 = 5700 - (3 \times 57) = 5700 - 171 = 5529$.

ASSOCIATIONS Although in Western society 100 is regarded as the common small round number, just as 40 is in Semitic literature (*see* discussion at **40**), "hundred" occurs fewer times than one might expect in common expressions—people often seem to prefer a thousand or a million to indicate many, as in the Chinese proverbs about a picture being worth a thousand words or a journey of a thousand miles beginning with a single step; or in the high-worth images of a million-dollar baby who flashes a million-dollar smile. While we round to hundreds, we express size with thousands and millions. The time interval of 100 years is common, however, and Franklin D. Roosevelt's administration famously began with 100 days of activity. Among well-known quotations using the word, "hundred" is followed by the word "years" in more than a third of the instances.

The relationship of 100 to virgin purity is on the authority of St. Jerome, who also informs us that 30 represents marriage and that 60 stands for widowhood. His reason for these associations comes from the common custom of the time (fifth century A.D.) of representing numbers and calculating with them using the fingers, nails, and knuckles of two hands. In this system the number 30 was symbolized by the somewhat difficult gesture of putting the nails of the index finger and thumb of the left hand together in what the various writers who describe the gesture insist is a "tender embrace" or words to that effect. The number 60 involves the same two fingers, but now the fingerprint side of the index finger is on the nail of the thumb. Jerome says that the index finger symbolizes the trouble and tribulation that press down on the widow, who is personified as the thumb. The same gesture on the right hand that means 60 on the left represents 100, which Jerome identifies with virginity only because it is on the other hand from marriage and widowhood.

Jerome's account of the meaning of these numbers emerges as a gloss on a passage in the New Testament known as the Parable of the Sower. Seed that falls on good ground may increase a hundredfold, sixtyfold, or thirtyfold. Jerome, best known for his translation of the Bible, wanted to explain why Jesus chose those three particular numbers (I am not convinced).

120   One Hundred Twenty
CXX   2 × 2 × 2 × 3 × 5

FIELD MARKS   The number 120 is the least *multiply perfect number* (read *multiply* as MUL-tuh-plee, not mul-tuh-PLY). A perfect number is one, like 6 or 28, for which the sum of its divisors, including 1 but not including itself, is equal to the number. A multiply perfect number is one for which the sum of the divisors, as defined, is equal to a multiple of the number. In the case of 120 the sum is 240, which is 2 × 120, so 120 is multiply perfect. Since the sum is twice the original number, 120 is multiply perfect of class 2.

Other recognizable traits: 120 is 5! (*see* **Factorial Family,** pp. 145–146) and part of the Pythagorean triple 119, 120, 169 (*see* **Pythagorean Family,** pp. 67–70).

SIMILAR SPECIES   The number 672 is the next multiply perfect number of class 2.

PERSONALITY   The product of any five consecutive integers is a multiple of 120. One of the integers out of any five in a row must be a multiple of 5, and others must be multiples of 4 and 3. This combination gives a factor of 60, but there are two even numbers in any four consecutive integers, so in addition to the factor of 4 there is a factor of 2. Put them all together and you get 120.

ASSOCIATIONS   The "great hundred" or "long hundred" of Germanic custom was 120 (*see* discussion at **12**).

121   One Hundred Twenty-One
CXXI   11 × 11

FIELD MARKS   The number 121 is the smallest number that is both a star number (*see* discussion at **13**), being 5$\star$, and also a perfect square, being 11^2. Furthermore, it can be expressed in the form $p^0 + p^1 + p^2 + p^3 + p^4$ since $11^2 = 3^0 + 3^1 + 3^2 + 3^3 + 3^4$. It is the only perfect square that can be expressed as a sum of consecutive powers of a natural number.

Other distinguishing characteristics: 121 is the least example of a number that is a three-digit palindrome. A palindrome is usually a word or statement that reads the same backward or forward, such as "radar" or "Madam, I'm Adam." A number palindrome is the same idea: for example, 123,454,321. It has been shown that for all two-digit numbers, reversing the digits and adding will lead to a palindrome, although the steps may have to be repeated. For example, the number 73 reversed is 37, so the steps to a palindrome are

$$73 + 37 = 110; \quad 110 + 011 = 121$$

so in this case there are two steps to a palindrome. The champion two-digit number is 89, which does not produce a palindrome until the twenty-fourth step. The resulting palindromic number is 8,813,200,023,188. Notice that 98 is a co-holder with 89 of this title.

SIMILAR SPECIES   The number 11 is the smallest nontrivial palindrome and the only two-digit number that can be a prime palindrome, since all other two-digit palindromes are multiples of 11. There are many three-digit prime palindromes, however—131, 151, 181, 191, 353, 373, 727, 757, 787, 797, 919, and 929.

PERSONALITY As the square of a prime, 121 is not very tractable for operations such as multiplication or division using methods based on factoring. But 121 is written in our numeration system with the two smallest digits so, as with 111, 112, 211, 212, and 222, it is easy to multiply, add, or subtract no matter what its factors are. Generally it is easier to combine doubling with the standard multiplication algorithm, however, to find quickly products involving 121 instead of simply writing out every step. For example, to find 347×121 you could first think 347 doubled is $700 - 6 = 694$, relying on the proximity of 347 to 350. Then the product is

$$
\begin{array}{r}
347 \\
694 \\
+347 \\
\hline
41987
\end{array}
$$

The point here is that the partial products can be found "in your head" instead of dealing with one digit at a time.

144 One Hundred Forty-Four
CXLIV $2 \times 2 \times 2 \times 2 \times 3 \times 3$

FIELD MARKS The number 144 is the only Fibonacci number (*see* **Fibonacci and Lucas Families,** pp. 120–122) that is also a perfect square. Oddly, if you count 1 as the first Fibonacci number so that the sequence begins 1, 1, 2, 3, 5 . . . , then 144 is the twelfth Fibonacci number as well as being the square of 12.

SIMILAR SPECIES The combinations of 2 and 3 repeated several times are easy to use but not always easy to recognize. Put another factor of 2 in 144 and you have 288. Another 3 brings 432, and one more takes you to 1296, the square of 36 and the fourth power of 6.

PERSONALITY Because it is composed of the two smallest primes each repeated, there are many helpful ways to factor 144. Depending on the computational need, you might use 2, 3, 4, 6, 8, 9, 12, 16, 18, 24, 36, 48, or 72 to unlock a problem involving 144.

ASSOCIATIONS Twelve dozen, or 144, is also known as a gross, a common unit of measurement for shipping or purchasing sets of small objects. The derivation of the word seems to be from an old word meaning "large" or "thick," which also is the original derivation of "gross" as coarse, meaning "disgusting" or "repulsive". But in medicine a "gross lesion" is not especially disgusting, just large enough to be seen without using a magnifying instrument.

---

196   One Hundred Ninety-Six
CXCVI   2 × 2 × 7 × 7

FIELD MARKS The number 196 emerges from the study of number palindromes (*see* discussion at **121**). Searches have failed to find a palindrome created by adding 196 to its reverse and repeating the process; this is the smallest number for which no amount of steps tested has produced a palindrome. No one has established whether or not a palindromic sum will eventually appear.

SIMILAR SPECIES 196 is the square of 14, which is somehow more easily overlooked than the more obvious squares around it: 144 and 169 just below and 225 and 256 just above.

PERSONALITY For 196, having two factors of 7 is somewhat redeemed by being close to 200, but it is often easier to multiply with half 196, which is 98, than with 196 itself. For example, to find 236×196 you would start with 236×98 and then break that down to $(100 - 2) \times 236$. Doubling 236 gives 472, so 236×98 is 472 less than 23,600 or 23,128. So 236×196 is twice that or 46,256 (a doubling made easier if you have memorized the powers of 2 so that you recognize 256 as twice 128 without doing any computation).

ASSOCIATIONS Another appearance of 196 is in conjunction with Bode's law, since 196 is the number in the sequence of Bode's law associated with the planet Uranus (*see* discussion at **28**).

---

220   Two Hundred Twenty
CCXX $2 \times 2 \times 5 \times 11$

FIELD MARKS The number 220 is one of a pair that are perhaps the most famous nonhuman couple of antiquity, the only two then-known *amicable numbers* (the other is 284). What makes these numbers more than simply friendly is that the sum of divisors less than the number itself but including 1 (as in the definitions of perfect, abundant, and deficient numbers) for 220 is 284 and the divisors of 284 sum to 220. For the record, the divisors of 220 are 1, 2, 4, 5, 10, 11, 20, 22, 44, 55, and 110.

SIMILAR SPECIES   Aside from 284, the next similar pair of amicable numbers were discovered by Fermat in 1636. They are 17,296 and 18,416. Furthermore, Fermat and Descartes worked out a method of finding at least some such pairs (Descartes gets the credit for 9,363,584 and 9,437,056). Euler and Lagrange went on to find more pairs. These began relatively low with the combination 2620 and 2924 and proceeded to large numbers of digits.

Thus, one of the most astonishing discoveries in mathematics occurred in 1867 when a sixteen-year-old Italian named B. Nicolò I. Paganini (no relation to the violin and guitar virtuoso) found that 1184 and 1210 are amicable.

365   Three Hundred Sixty-Five
CCCLXV   5 × 73

FIELD MARKS   The number 365 is the smallest number that can be expressed as both the sum of two consecutive perfect squares and the sum of three consecutive perfect squares: specifically, $13^2 + 14^2 = 10^2 + 11^2 + 12^2 = 365$.

SIMILAR SPECIES   After 365, the next number that is the sum of both two and three consecutive perfect squares is $35{,}645 = 133^2 + 134^2 = 108^2 + 109^2 + 110^2$. After that comes 3,492,725, which I leave for you to decompose.

PERSONALITY   The difficulty of computing with a product of two primes that is not near any friendly number cannot be alleviated even by working with twice 365, which is 730. It may sometimes be handy to compute the number of days in a particular number of years, keeping leap year in mind. For example, a U.S. president's term of four years is typically 1461 days; two terms is 2922 days.

168

ASSOCIATIONS The number 365 is familiar to one and all as the number of days in a year. The statue of the Roman god Janus in the Roman Forum, according to Pliny the Elder, looked forward and backward (he was the god of gates and doorways), and his hands featured the number 365 in classical finger notation, to symbolize the year. January was the month of Janus, which the Romans in 154 B.C. changed to become the beginning of the year from its previous start on March 1. Why change? There was a rebellion in Spain. To speed up a change of commanders, normally an operation taking place at the turn of the year, 154 was given only ten months and the year we call 153 began on January 1. Thus, the two-faced Janus came to symbolize looking forward to the new year and backward to the old. Pliny's description of the statue was written some two hundred years after the change of the start of the year, so there was plenty of time for a tradition to develop.

1001   One Thousand One   MI   7 × 11 × 13

PERSONALITY Multiplying by 1001 is exceptionally easy if there are fewer than four digits in the other factor. For a single-digit factor, d, the product is always written $d00d$ (where d is treated as a sign instead of as a number). If the number is shown with two digits as de, product $de \times 1001$ is $de0de$, while the three-digit version shows $def \times 1001 = defdef$. When there are four or more digits, you actually have to add a bit to get the product, although it is still easy.

ASSOCIATIONS A number game is based on the idea that multiplying a three-digit number by 1001 is found by writing the digits down twice, a process that is not obviously multiplication. If you tell someone to think of a three-digit number and write the digits down twice (thinking, say, of 531 and writing 531531) to get a new number, the resulting multiplication is hidden. Now you can ask the sucker to divide this number by some low primes, steering your mark to 7, 11, and 13 (in any order) with warnings such as "no, not 17, that's too hard" or "not 5, that's too obvious." If the divisions are by 7, 11, and 13 and no other numbers, the final quotient will be the number that the sucker chose in the beginning—magic of a sort.

The idea of excess is often carried by adding "and 1" to a larger number, in particular by 1001, as in Scheherazade's tales of the 1001 nights.

1024   One Thousand Twenty-Four   MXXIV
$2 \times 2 \times 2 \times 2 \times 2 \times 2 \times 2 \times 2 \times 2 \times 2$

FIELD MARKS   The number 1024 is 2^{10} and as a result occurs in many problems.

SIMILAR SPECIES   The next higher tenth power is 3^{10}, which is 59,049, demonstrating how rapidly powers increase.

PERSONALITY   Computation involving several powers of 2 is much easier to do with the rules for exponents than by using the decimal expansions. For example, to divide 1024 by 64 think $2^{10} \div 2^6 = 2^{10-6} = 2^4 = 16$.

ASSOCIATIONS   When the prefix *kilo-* is used in computer terminology, it means 1024, not 1000 (*see* **Powers of 2,** p. 171).

Powers of 2

Numbers such as 2, 4, 8, 16, 32, 64, 128, 256, 512, 1024, 2048, 4096, 8192, 16,384, 32,768, and 65,536 occur whenever there are only two possibilities to consider. They appear in counting problems that can be analyzed as a sequence of decisions that can be called *choose* or *reject*. Perhaps the powers of 2 are most familiar to people today as the numbers that occur repeatedly whenever computers are discussed, but they are usually in disguise. Although the prefix *kilo-* means 1000, a kilobyte is not 1000 bytes but 1024 bytes. This is because the computer makes a series of choices in the choose-or-reject format, commonly labeled *on* or *off*. Thus, if there are ten such choices to be made, the number of possible states for the computer is $2^{10} = 1024$, conveniently close to 1000 but not equal to it. As a result, adding 32 gigabytes of memory to your computer is adding rather more than 32,000,000,000 bytes (*giga-* means 1 billion). Instead, you are adding $32 \times 2^{30} = 2^{35} = 34,359,738,368$ bytes.

10^{63} The Sand Reckoner's Number

FIELD MARKS   In principle the Hindu-Arabic system can be used to write any natural number, no matter how large. The Greeks did not have such a system, and so it was to some degree debatable among them as to how far numbers might reach. While one school of philosophy proposed paradoxes that involved the possibility and impossibility of infinite numbers, other philosophers could cite Aristotle, who proposed that true infinity could never be realized. The common people seem to have thought of infinity as merely a very large number (if they thought of it at all).

Into this situation stepped Archimedes of Syracuse [c. 287–212 B.C.], generally recognized as the greatest all-round mathematician of classical times. His discussion of large numbers, *The Sand Reckoner*, begins with the following introduction:

> There are some . . . who think that the number of the sand is infinite in multitude; and I mean by the sand not only that which exists about Syracuse and the rest of Sicily, but also that which is found in every region whether inhabited or uninhabited. Again there are some who, without regarding it as infinite, yet think that no number has been named which is great enough to exceed its multitude. . . . But I will try to show . . . of the numbers named by me . . . some exceed a mass [of sand] equal in magnitude to the universe.

Archimedes then proceeded to compute the volume of the universe based on the premise, put forward a few years earlier by Aristarchus of Samos, that Earth revolves around the Sun and that the celestial sphere of stars is far beyond the orbit of Earth. Archimedes was very careful to say that this strange idea of Aristarchus is just a hypothesis, and he rejected as illogical the suggestion from Aristarchus that the ratio of the distance to the stars to Earth's orbit must be as the diameter of a circle to its center (a nice approximation of what we now know to be the truth if you don't mind dividing by 0, which Archimedes was unwilling to do). But Archimedes began with the idea that the stars are as far from Earth as the ratio of the orbit of Earth is to its diameter. In making his calculation, he deliberately overestimated the size of Earth by a factor of 10 (basing his calculations on then current ideas and measuring in the common Greek measure of stadium lengths) and also provided good margins of error on the distances involved, explaining each one carefully. In estimating the volume of grains of sand, for which he errs on the other side, he began with 10,000 grains per poppy seed, 40 poppy seeds measuring the breadth of a finger, and the length of a stadium equal to less than 10,000 finger breadths. When he was done he had every linear measure converted to stadium lengths. From this result, he could calculate the volume first, and then return to number.

Before he could complete his argument, however, Archimedes had to invent a new notation system, and the one that he created is far more powerful than needed to represent the number of grains of sand. At that time the largest number available in the Greek system was the myriad, which is 10,000 in our system. Archimedes proceeded to designate what he called the *first order* (that is, the first order after the myriad) the "myriad myriads," which we would write as 100,000,000. What he called the *second order* uses that myriad myriads as its unit, so after another myriad myriads, the numeration system has reached what we would write as $100,000,000^2$, or the original myriad raised to the fourth power. Archimedes then continued to go order after order until reaching the *myriad myriadth order*, for which the highest number is $100,000,000^{100,000,000}$, which is known as P. The number P might be more recognizable as $10^{800,000,000}$.

This vast number of orders completes only the first *period* in Archimedes' system. Then he goes through a series of periods, winding up with the *myriad myriadth period*. The last number of the last period is $10^{80,000,000,000,000,000}$, or 10 to the 80 thousand trillionth power.

This was more than sufficient for his original purpose. The number of grains of sand in a universe, built more or less according to Aristarchus' radical idea that Earth revolves about the Sun, would be less than the number we would write as 10^{63}.

SIMILAR SPECIES The largest number encountered by most Americans in the course of their daily lives is the number termed the national debt. At the end of the twentieth century, the national debt of the United States was running at about \$6 trillion. As a number in decimal notation that is 6,000,000,000,000, or in scientific notation (*see* below) 6×10^{12}. So the national debt is less than the number of grains of sand in Aristarchus' universe. Another familiar number, the number of people on Earth, just reached 6,000,000,000 (one thousandth the U.S. national debt) in 1999—much smaller than the Sand Reckoner's Number.

Today the universe is a little bit bigger, and we know of entities much smaller than a grain of sand. Estimates vary, but a typical view of the total number of subatomic particles in the Hubble Space Telescope view of the universe is 10^{76}. This does not seem at first to be much larger than Archimedes' number, but it is a bigger difference than is obvious at first sight. Comparisons for numbers shown in the form 10^n are particularly easy to make—you need only subtract the numbers represented by the exponents to find by what power of 10 the lower number needs to be multiplied to equal the higher.

Here is a homely example. How much bigger, as a ratio, is a billion than a million? A billion in the United States at least is 1,000,000,000, or 10^9, while a million is 1,000,000, or 10^6. So an American billion is $10^{9-6} = 10^3$ times as big as a million. (In England the billion is a different number, 10^{12}, which is a trillion in the United States, so an English billion is 10^6 times or a million times as big as a million. The French go with the American system, the Germans with the

English. The main problem comes in books that hope to sell in both England and the United States, causing the authors to write 1,000,000,000 instead of a billion or—in England—milliard and 1,000,000,000,000 instead of a trillion or billion.) Thus, to compare 10^{76} with 10^{63} think $76 - 63 = 13$, so 10^{76} is 10^{13} times as large as 10^{63}. In other terms, a presently accepted estimate for the number of subatomic particles in the universe is 10 trillion times as large as the Sand Reckoner's Number.

PERSONALITY Scientists often speak in terms of "orders of magnitude," an expression that is seldom defined explicitly despite its common usage (*see* discussion at **10**). If one number is an order of magnitude greater than another, then the first is ten times the latter. This idea works especially well with the system of notation based on powers of 10 that is known as *scientific notation*. This notation is useful for the extremely large and small quantities encountered in science and to indicate the precision of a measurement. When we say that the Sand Reckoner's Number is 10^{63} we are using scientific notation; in ordinary Hindu-Arabic notation the same number would be

1,000,000,000,000,000,000,000,000,000,000,000,
000,000,000,000,000,000,000,000,000,000

which cannot fit on a single line of type, since there are sixty-three 0s following the 1.

But scientific notation would be rather special if it could be used only for powers of 10. Instead, each number is expressed as a decimal fraction that is between 1 and 10 (the multiplier) times an integral power of 10. By keeping the multiplier between 1 and 10, the exponent for the power of 10 tells the size. For example, the number shown in scientific notation as 1.2334×10^9 will be about a billion, an order of magnitude greater than the number 1.2334×10^8. It should be clear that this system, which combines ordinary decimal notation with powers of 10, can be used to express any number whatsoever. If the number is actually a power of 10, the multiplier is omitted. If the number is less than 1, a negative power of 10 is used (*see* discussion of *rational exponents* at **Genus Rational, 1/4**).

One of the many virtues of scientific notation is that computation with numbers expressed in scientific notation is simplified, since $(a \times 10^b) \times (c \times 10^d)$ is the same number as $(a \times c) \times 10^{b+d}$. Similarly, $(a \times 10^b) \div (c \times 10^d)$ is just $(a \div c) \times 10^{b-d}$.

Another virtue of scientific notation is that the decimal multiplier is supposed to show exactly how a measurement was carried out. For example, as a pure amount the numbers 1.2×10^3 and 1.200×10^3 are both the same as 1200. But the number of decimal places should indicate the precision of the measurement. When that convention is used, it is understood that a measurement of 1.2×10^3 meters was measured to the nearest 100 meters, but one of 1.200×10^3 was measured to the nearest meter.

10^{96} Rotman's Biggest Number

FIELD MARKS According to Brian Rotman of Louisiana State University at Baton Rouge, 10^{96} is the largest possible number that could be reached by counting using a finite computer operating at a theoretical minimum drain on energy. The effort needed to count this high would exhaust all the energy in the universe, including the dark matter that cannot be detected.

It is Rotman's notion that just as there is a non-Euclidean geometry in which no lines are parallel, such as the system in which a straight line is a great circle on a sphere, there is a non-Euclidean arithmetic in which counting does not extend to infinity. Rotman notes that completed infinities (*see* discussion at **Kingdom Infinity, ∞**) do not exist in nature, so why should they in mathematics? Counting, the process of iterating one number after another, can go only so far. Rotman figures that counting is limited practically by how high a number can possibly be produced by counting. Since the computer can be programmed to count very fast, he considers the highest number that a theoretically possible computer might reach, which would be limited by energy requirements. Rotman calls numbers higher than that (theoretically) highest possible number, which is 10^{96}, "transiterates" instead of numbers. Transiterates might not behave at all like the numbers that can actually be counted. Note that transiterates are not infinities, but "numbers higher than you can count to."

SIMILAR SPECIES Rotman's Biggest Number, 10^{96}, is not quite as big as a googol, which is 10^{100}, or 1 followed by a hundred zeroes (*see* 10^{100}). The difference between the googol and Rotman's Biggest Number is that no entity in the universe could possible count to a googol, which is 10,000 times as big as Rotman's Biggest Number. To Rotman, a googol is merely a fairly small transiterate.

Rotman suggests that the largest number one could even imagine counting to by a computer the size of the entire universe would be 1 followed by 10^{98} zeros, which might be called Rotman's Even Bigger Biggest Number.

PERSONALITY Perhaps we forget to use our mental number line when confronted with numbers in exponential form. It is easier to recognize that 1234 is about a hundred times as big as 12 than it is to observe that 10^4 is exactly a hundred times 10^2. My suspicion is that we learn to associate the length of the physical numeral with the size of the number, so that 10^{96}, even if it is the largest number that a perfectly efficient finite computer could count, does not seem nearly as much as 1,234,567,890,987,654,321, which is just a little bit more than a sextillion.

ASSOCIATIONS   While Rotman considers the theoretical border line between the transiterates and what, on the basis of what a computer can do, he seems to think of as "actual numbers," it might make more sense to put this on a human scale. As everyone who has measured the time in seconds by counting "a thousand and one, a thousand and two, a thousand and three," and so forth will recognize, a person can easily count aloud more than a number name per second up to about 1000; and after that for a time the rate will be about a number name pronounced per second, while some large numbers ("a billion two hundred thirty-seven million five hundred eighty-nine thousand seven hundred ninety-seven" for example) take more than a second to name even if you hurry. Consider someone who learns to count at about the age of four and lives to be a hundred and four. How high will the count be at the hour of death? In this counting lifetime there are going to be 100 times 365 days plus extra days for leap years (24 leap years, most likely, since there is no leap day in most years that end with two zeros), so the number of days is 36,524. Each of those days contains 24 hours, for a counting life of 876,576 hours, each hour filled with the counting of 60 × 60 = 3600 number names. Multiply the total hours by 3600, and you hear "3,155,673,600" whispered on a dying breath. Thus, in human terms, a number of about 3 billion would seem to be the largest that might be meaningful, if you accept the idea that no number is meaningful unless it can be reached by counting.

10^{100} The Googol

FIELD MARKS   The famous googol was originally defined by its numeral as 1 with a hundred zeros following it (10^{100}), which was the largest number that could be described (at first) by a bright nine-year-old boy, the nephew of the mathematician and mathematics popularizer Edward Kasner [American, 1878–1955] in the late 1930s. The googol has remained a part of the popular American numeration system, appearing in crossword-puzzle dictionaries, almanacs, and even in standard dictionaries. Mathematicians rarely use it, but scientists sometimes might say "googol" as a handy way to refer to a very large number.

SIMILAR SPECIES   In the book he wrote with James R. Newman, *Mathematics and the Imagination*, Kasner wanted to point out that very large numbers are different from infinity, and he reported on his experiences in working on very large numbers with kindergarten children. Assisted by Kasner, the children determined that the number of raindrops falling on New York City in a century was less than a googol, and Kasner calculated that the number of grains of sand on Coney Island Beach was about 10^{20}. Kasner also calculated the number of words spoken by all the humans in history (to 1940, of course) as 10^{16}, a number that he also assigned to all the words ever printed. He mentions that as a result of inflation in the 1920s, the Germans had $4.96585346 \times 10^{20}$ marks in circulation—during this period some housewives famously used wheelbarrows to carry their bills when shopping for groceries.

The googol is sometimes illustrated in terms of time or distance. If we take the universe to be about 15 billion years old, which is on the high side of current estimates, then fewer than 10^{19} seconds have elapsed. Consider a shorter time interval: the second is officially defined as 9,192,631,770 ticks of a cesium atomic clock, but even such a clock would have ticked fewer than 10^{28} times since the big bang at the start of the universe. The distance in miles from Earth to the Sun averages slightly less than 10^8 miles, a distance that is somewhat less than 10^{13} inches. If we assume that the distance from one side of the universe to another is 30 billion light-years, and use the approximation 5,580,000,000,000 for the number of miles in a light-year, then the diameter of the universe is about 10^{26} miles or 10^{31} inches, while the circumference does not even come to 10^{32} inches.

$2^{6,972,593} - 1$ Largest Prime Known (1999)

FIELD MARKS   As of 1999, the largest known prime number was $2^{6,972,593} - 1$. The number was located as part of the Great Internet Prime Search, a cooperative project that anyone with a moderately powerful computer can join (Web site at http://www.mersenne.org/). The numeral for this large prime has 2,098,960 digits when written in Hindu-Arabic decimal notation instead of the exponential form. This size may not seem like a lot in the context of this discussion of a whole group of very large numbers, but think of it this way: the average word has about four or five letters, depending on the level of difficulty of the text—a romance novel might have four-letter words on the average and a treatise on quantum physics might have five. So if all the digits in the largest prime were printed as "words" containing four digits with spaces between them, the digits would occupy the same space as 419,792 short words, about the length of seven romance novels.

SIMILAR SPECIES

The number $2^{6,972,593} - 1$ is written in this somewhat peculiar form because it is a type of number familiar to mathematicians as a Mersenne prime (*see also* **Mersenne Family,** pp. 85–87). Numbers of the form $2^p - 1$ where p is a prime number are called Mersenne numbers, and if the resulting number is also prime, then $2^p - 1$ is a Mersenne *prime*. (Some writers call a number a Mersenne number only if it is also prime.) Mathematicians have found special methods for testing any Mersenne number to determine whether or not it is prime. Nevertheless, $2^{6,972,593} - 1$ is so large that it took 111 days of calculation on a desktop computer to determine that it was prime and not composite, although if the computer had done nothing else during that time it could have been checked in about three weeks. The number $2^{6,972,593} - 1$ is only the thirty-eighth Mersenne prime found, although it may not be the thirty-eighth Mersenne prime: not all candidates less than $2^{6,972,593} - 1$ have been checked.

It may seem peculiar that anyone would be especially interested in numbers of the form $2^p - 1$, but fascination with such numbers has a tradition in mathematics that traces back to Euclid. Euclid's theorem on perfect numbers (*see* discussion at **6**) states that all numbers formed as the product of 2^{n-1} and $2^n - 1$ are perfect numbers provided that $2^n - 1$ is prime. The first few numbers n that you might examine are $n = 0, 1, 2, 3, 4, 5$. When $n = 0$, $2^n - 1$ becomes $1 - 1 = 0$, so that gets you nowhere. Similarly, when $n = 1$, the number expressed by $2^n - 1$ is also 1; thus, $n = 1$ is excluded from consideration by Euclid's theorem because 1 is not considered a prime.

The first allowable instance is for $n = 2$, in which case $2^n - 1$ is 3, which is prime. Then Euclid's theorem states that the product of $2^{2-1} = 2$ and $2^2 - 1 = 3$ will be a perfect number, and indeed 6 is perfect. Similarly, when $n = 3$ the number $2^n - 1$ is 7, which is prime, the number 2^{3-1} is 4, and the Euclidean product is $4 \times 7 = 28$, which is also perfect. But when one looks at $n = 4$, the number $2^n - 1$ is 15, which is not prime, so Euclid's theorem does not apply. Next up is $n = 5$, which produces 31, another prime, so Euclid's theorem again applies. The other factor is $2^{5-1} = 16$, and the product is 16×31, which is 496, yet another perfect number.

It cannot escape notice that in these calculations $2^n - 1$ was prime when n was also prime, which might suggest that this will always be the case. Mersenne, however, concluded from his investigations that the only primes p for which $2^p - 1$ is also prime are 2, 3, 5, 7, 13, 17, 19, 31, 67, 127, and 257. It was this assertion that made Mersenne eponymous, although in the course of mathematical investigation Mersenne was proved wrong. But Mersenne was correct in observing that not all numbers of the form $2^p - 1$ are prime. For example, if p is 11, the number $2^{11} - 1 = 2047$, which can be factored as 23×89.

Before the days of computers, calculations of Mersenne numbers and tests of whether they are prime or composite were very difficult. Mersenne himself overlooked $p = 61$, 89, and 107, all of which are less than the highest numbers of his conjecture.

PERSONALITY Pierre Fermat in June 1640 wrote to Mersenne about some of the properties of $2^n - 1$ that he had unearthed. Fermat showed that if n is not prime, then neither can $2^n - 1$ be prime, and also that $2^p - 1$, where p is prime, can be divisible only by numbers of the form $2kp + 1$, where k is a natural number.

ASSOCIATIONS   Although $2^{6,972,593} - 1$ is the largest prime recognized as of 1999 (discovered on June 1), it is certainly not the largest prime that exists and very likely not the largest prime that will be known by the year 2000. It seems that since computers have become common that someone encounters a prime larger than any previously known about once every three years, although I have not done a careful study of this. The pace may be picking up, since "largest primes" were found in 1996, 1997, and 1998 as well as 1999.

The reason that one can say with certainty that $2^{6,972,593} - 1$ is not the largest prime is that more than 2000 years ago in one of the most significant proofs of mathematics Euclid proved that there is no largest prime. Here is a brief sketch of Euclid's Proposition 20 from Book IX: prime numbers are more than any assigned multitude of prime numbers.

Consider that the number N is the largest prime there is. Form the product of all the primes less than N, which is surely a natural number, which we can call P. Note that P is greater than N, the largest prime. Now consider an even larger number, $P + 1$. If $P + 1$ is prime, the theorem is proved, since there is a prime greater than N. But if $P + 1$ is not prime, then it has a divisor that is a prime. Could that divisor be any of the primes less than N? No, because each of these when divided into $P + 1$ will leave a remainder of 1. Therefore, if there is a number that divides $P + 1$, that number must be larger than N. Hence, N is not, as assumed, the largest prime.

$10^{10^{98}}$ Rotman's Even Bigger Biggest Number

FIELD MARKS   According to Brian Rotman (*see* discussion at 10^{96}), $10^{10^{98}}$ is the largest "practical" number possible. If one designed a computer that was as large as the entire universe, a computer whose sole job is to store numbers from 1 to as high as one can go, this would be the largest storable number. Under his prescribed conditions, the universe-sized computer having stored from 1 to the number that is 1 less than

$$10^{10^{98}}$$

would then require all the energy in the universe to store one more number.

SIMILAR SPECIES   The number 10^{96} is calculated by Rotman to be the highest number that an ideal computer that is less than the size of the entire universe, but which operates using the theoretical minimum of energy, could count to before it used all the mass-energy in the universe.

A number that is sometimes put forward as the number of combinations of genes possible for humans is $10^{2,400,000,000}$, a determination based on the theory of counting arrangements of sets. This number, like many of the other large numbers from science, is based on some assumptions that have yet to be pinned down as facts. No one at this time knows the number of genes in the average human, and I have seen estimates that range from 50,000 to 200,000.

Even the smallest possible number of genes would allow for a large number of combinations. With some exceptions, each person inherits at random one of each pair of genes from the mother and one from the father (some genes from

184   Kingdom Number

the mother are unpaired in male offspring, and there may be other instances related to sex of unpaired genes). The normal way of calculating this kind of choice is that for the first gene pair there are two choices for each gene in the pair—the mother's gene or the father's. The possible combinations are MM, MF, FM, and FF (that is, both from the mother, two different ways to get one gene from the mother and one from the father, and both from the father). Each of these combinations might be paired with one of the four similar combinations for the second gene pair, producing $4^2 = 16$ possible combinations from two gene pairs. Add another gene pair and each of the sixteen from the two pairs can combine with one of the four possibilities for the pair, bringing the number to $4^3 = 64$ combinations. As you can see, if this analysis accounts for all the variation, the total number of combinations would be somewhere between $4^{50,000}$ and $4^{200,000}$, depending on how many gene pairs there are. But even $4^{200,000}$ is far short of $10^{2,400,000}$. My spreadsheet program won't calculate anything larger than 4^{511} (don't ask me why 511 is OK but not 512—my guess is that I must be running out of registers on my 32-bit computer). From this computer calculation, however, I can observe that 4^{511} is not quite as much as 10^{308}.

But the biology is far more involved than just four possible combinations for each gene. That is because humans have many different possibilities, or alleles, for each gene. So the mother's gene for a particular trait might come in as many as 200 different forms, while the father's gene for that trait might also come in the same number of different forms. Unless one has some estimate of the number of possible variations, it is hard to see how the correct number of possibilities can be counted. Furthermore, some combinations of nonlethal genes could be lethal, reducing the number of viable humans. In any case, the number of humans who have lived since the beginning of our species is probably less than 20 billion and therefore certainly fewer than 10^{10}. So even if there were only 10^{308} possible combinations, we would be a long way from running out of different people.

Another frequently cited number that is somewhere in the vicinity of Rotman's Even Bigger Biggest Number is the number of possible moves in a chess game. This is calculated to be

$$10^{10^{50}}$$

It might be tempting to think that this is about half Rotman's Even Bigger Biggest Number since 50 is about half 98, but that would be far from the mark. If you think of this number as 1 followed by 10^{50} 0s and the larger Rotman number as 1 followed by 10^{98} 0s, then you can calculate by subtracting 50 from 98 that the Rotman number is 1-followed-by-10^{48} 0s times as large as the number of chess moves.

Surprisingly, however, Rotman's Even Bigger Biggest Number is just slightly *smaller* than a number that was invented to be as big a number as anyone might think of. This number is the googolplex, which is 1 followed by a googol of zeros (*see* 10^{googol}). Although the googolplex was invented to be as large a number as one might think of, it is clearly nowhere near as large as 1 followed by a googolplex of zeros, a number so great that it fails even to have a name.

For many years the largest number with which anyone dealt was Skewes's number, which occurs in a mathematical proof. It is usually given as

$$10^{10^{10^{34}}}$$

which, since it is the same as

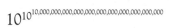
can easily be seen to be much larger than Rotman's Even Bigger Biggest Number. A more recent version of Skewes's proof, however, has reduced the number needed to make the proof work to a mere 10^{1167}, which is much smaller than Rotman's Even Bigger Biggest Number, although larger than Rotman's Biggest Number.

PERSONALITY   It is difficult to imagine a number as great as this. Since 10^{96} is Rotman's proposed largest number that is even theoretically possible, this number, 1 followed by 10^{98} zeros, has a hundred times as many zeros as a finite, less-than-universe-sized computer could even count.

10^{googol} $10^{10^{100}}$ The Googolplex

FIELD MARKS   The googolplex, like the googol (*see* 10^{100}), was proposed by Edward Kasner's nine-year-old nephew in the 1940s as a number even larger than the (previously) largest number he could imagine. It remains the largest number that is occasionally mentioned by writers about mathematics, most often to show how even very large numbers can be expressed compactly or to provide a readily comprehended example of a very large number.

SIMILAR SPECIES   Notice that the googolplex is even somewhat larger than Rotman's Even Bigger Biggest Number, so that no imaginable process could count it out.

Genera *Integral* and *Rational* (Signed Numbers and Fractions)

Genus *Integral*

The German mathematician Leopold Kronecker [1823–1891] is reported to have said, "God made the integers, and all the rest is the work of man." But he was probably not thinking of the integers as we know them in English, positive and negative, but of what we call the natural, or counting, numbers. Some mathematicians even as late as the nineteenth century have found the negative branch of the genus *Integral* unacceptable to humans, much less to God.

The integers are the natural numbers and their opposites. In a formal way, the opposites are sometimes defined by saying that if a is a natural number, then its opposite is a number $-a$ such that the sum of a and $-a$ is 0. There is also a more intuitive definition: an opposite to a is a number $-a$ that is as far from 0 as is a, but in the opposite direction. The definition using 0 as a sum might be called the accountant's definition, while that using 0 as a point from which to measure might be termed the thermometer definition.

Historically, the accountant's definition appears to have precedence. Ancient Chinese scribes used a system of negative numbers to record debts. Chinese numerals come in two varieties, one of which descends by a direct line from the use of bamboo, ivory, or iron rods as counting devices used for calculations (the abacus was invented later.) The rods were of two types, red and black. Although it would be tempting to think that red represented negative numbers, just as a modern accounting program will put numerals for negatives in red type, remember that this is China, where the color of mourning is white, not black. In China red is the color of good luck and prosperity. Thus, the Chinese scribes used red rods for positive numbers and black ones for negatives. Later they used written numerals of the same colors to show negative and positive. Indian mathematicians and accountants learned a lot from the Chinese, and also used negative numbers to show debts and positive ones to show assets.

About A.D. 628 the mathematician Brahmagupta [Indian, c. 698–c. 665] stated for the first time in the record of mathematical history the rules of operation for negative numbers. Addition presents no special problem when negatives are interpreted as debts, but the other operations tend to be somewhat counterintuitive. Subtraction can be confusing, while multiplication and division partake of one rule that we have come to accept in English grammar—a double negative would be a positive—but for which mathematics teachers have invoked such tortured techniques as showing a film of people walking backward and then running it in reverse (to show people apparently, and awkwardly, walking forward). All this to explain why negative numbers have the odd-seeming property that the product or quotient of two of them is positive. Some just give up trying to explain the peculiar nature of multiplication of negatives. W. H. Auden tells that in his British schooldays they learned the rule as "Minus times minus equals plus. / The reason for this we need not discuss."

Another discovery of the Indian mathematicians, now familiar to every high school student, is that positive numbers all have two square roots, one positive and one negative. On the other hand, negative numbers themselves have 0 square roots among the real numbers, while 0 has just one root, since $\sqrt{0} = 0$, which is neither positive nor negative.

Negative numbers were not easily accepted in the West, however. René Descartes was willing to allow numbers less than 0 only because he found ways to convert problems with negative solutions into problems with positive ones (so it is OK to report a negative answer, but to make sense of it you need to change the question). Almost 200 years later the logician Augustus De Morgan [English, born in India; 1806–1871] substantially agreed, but stopped short of allowing the original negative answer to make any sense.

Blaise Pascal was another seventeenth-century mathematician who rejected the whole notion. One of his friends, Antoine Arnauld [French, 1612–1694], used the idea of ratio to attack the negatives. He said that according to the division rules for negatives, 1 is to -1 as -1 is to 1, but since -1 is less than 1, this statement says that a lesser number is to a greater as a greater one is to a lesser. Can't be. Many mathematicians of the time argued about this one to no avail. The German philosopher and mathematician Gottfried Wilhelm von Leibniz [1646–1716] finally ended the discussion by saying, in effect, that the philosophy may not make sense but it is more important that computation with negative numbers works, so forget it.

The English mathematician and theologian John Wallis [1616–1703], whose work influenced Newton, thought that negative numbers are greater than infinity, so he had no problem with Arnauld's ratio argument. This view may seem odd to the modern reader. Wallis based it on the common idea, also incorrect, that $a/0 = \infty$. He argued that a/b grew larger and larger as b grew closer to 0, becoming infinite when $b = 0$, so beyond 0 the negatives must be greater than infinity to reduce the size of a/b when b is negative; or perhaps he believed that the quotient a/b is greater than infinity when b is negative. Euler, over a hundred years later, also thought that negative numbers are greater than infinity. In these cases, it appears that the mathematicians believed in the existence of a completed infinity (*see* discussion at **Kingdom Infinity,** $\aleph_0$), although perhaps they simply believed that the negatives were greater than any of the natural numbers (which would not necessarily require a completed infinity).

In some cases, negative numbers partake of the properties of their positive counterparts. For example, when the concepts of "prime" and "composite" are extended to include the negatives of the counting numbers, we can specifically say that "any number is prime if its only factors are 1 or -1 and its absolute value," which makes numbers such as -2, -3, -5, -7, and -11 prime, while numbers such as -4, -6, -8, -9, -10, and -12 are composite. On the other hand, some properties of counting numbers are hard to extend to the negatives. You could define -1, -3, -6, -10, and -15 as triangular numbers, for example, by saying that the negatives of the triangular numbers are also triangular, but it is difficult to see where that would lead. There seems to be no sensible image of these "negative triangular numbers" nor any use for them. A more extreme case concerns the square numbers, since such numbers as -1, -4, -9, -16, and -25 are clearly not the squares of negative numbers.

A common way for mathematicians to contemplate almost any number system after the counting numbers is in terms of ordered pairs, which can be done for the integers and also for the other genera of numbers that we discuss later. The virtue of ordered-pair notation to mathematicians is that ordered pairs of whole numbers (or, as we shall see, of real numbers) fit easily into axiomatic development of a number system. For integers, there are two types of ordered pairs, corresponding to the familiar positive and negative numbers. The ordered pairs $(1, 0)$, $(2, 0)$, $(3, 0)$, and so on, are taken to be the natural numbers with the underlying understanding that the notation $(1, 0)$ really stands for the set $\{1, \{0\}\}$. Of course, $(0, 0)$ is the identity element (*see* **Genus Natural,** 1) in this system. Then the negative numbers are $(0, 1)$, $(0, 2)$, $(0, 3)$, and so on. Equality is defined in the usual way, which is that $(a, b) = (c, d)$ if and only if $a = c$ and $b = d$. If both addends are positive or if both are negative, then addition is also defined in the usual way, so that $(a, 0) + (c, 0) = (a + c, 0)$ while $(0, b) + (0, d) = (0, b + d)$. When one addend is negative and the other positive, it is necessary to know which counting number is the greater. To add $(a, 0) + (0, b)$ if $a > b$ then the sum is $(a - b, 0)$, but if $b > a$ then the sum is $(0, b - a)$. Multiplication also results in some definitions that are expected and some that are more difficult to anticipate. The operation we might expect is $(a, 0) \times (c, 0) = (a \times c, 0)$, since the rule is the same as for the natural numbers. But $(0, b) \times (0, d) = (b \times d, 0)$; this is the rule for integers that the product of two negatives is positive. Finally, for $(a, 0) \times (0, b)$ the product is $(0, a \times b)$, corresponding to the rule that the product of two numbers of opposite sign is always negative.

Turning the familiar rules for operations with integers into their ordered-pair version may seem to some like a useless exercise. But it is very Pythagorean, since it shows that everything that we normally do by adding a new genera of numbers to the taxonomy could be done with just the counting numbers and 0.

Commonly Seen Species

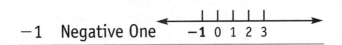

−1 Negative One

FIELD MARKS Multiplication by −1 changes every number but 0 to its opposite, so −1 is both a number and the operator "taking the opposite." An *operator* in mathematics or mathematical physics is any entity that is a specific unary operation. Operations such as addition and multiplication, which produce a third number by combining two numbers, are binary operations. But many common operations are of the unary sort, producing a second number directly from one number. Some familiar examples of unary operations are taking the (positive) square root, taking the absolute value, and taking the logarithm. The unary operation, or operator, "taking the opposite" is exactly the same as the binary operation "multiplying by −1."

SIMILAR SPECIES The number $i = \sqrt{-1}$ is related in several ways to −1, but while i is both a number and an operator, it is not the same kind of operator as −1. Multiplication by −1 moves a number from its position on the number line to a position that is the same distance from 0 (same absolute value) in the opposite direction. Multiplication by i, on the other hand, moves the number from its position on a number line to the same position on a line perpendicular to the original line.

An even better comparison for the number −1 would be to an operator that is not even in Kingdom Number, but is from a different mathematical realm, that of the transformations. *Inversion* is a transformation with respect to a circle that transforms each point in space that is outside the circle to one inside and each inside point to one outside. In a famous sequence of ways that a mathematician might capture a lion in the desert—all parodies of the behavior of real mathematicians in dealing with mathematical objects—my favorite has always been capture through inversion. The mathematician, placed inside a cage, then performs an inversion transformation, after which everything outside the cage, including the lion, is inside the cage and the mathematician is outside.

PERSONALITY   Beginning algebra students often have difficulty in dealing with problems that include -1 hidden in one way or another. It helps if they first learn to abandon the operation of subtraction completely, since it only causes confusion. Algebra, in theory at least, replaces subtraction entirely by the definition that subtraction is addition of the opposite, so an expression such as $x - 2$, while often read as "ex minus two," is considered to be an abbreviation for $x + (-2)$ and is handled by the rules of addition. Since multiplication by -1 is the same operator as taking the opposite, an apparent subtraction problem such as $3x - 5x$ is taken to be a shorthand way of writing $3x + (-1)5x$. In such a simple example there is usually no confusion, but when a problem becomes somewhat more complicated, such as "subtract $3x - 5y$ from $2x - y$," rather more difficulty is encountered. This can be greatly eased if a person learns to think "subtract $3x - 5y$ from $2x - y$; that means add $(-1)(3x - 5y)$, or $-3x + 5y$, to $2x - y$, so the answer is $-x + 4y$." A related problem occurs when an expression with several terms appears within parentheses and preceded by the subtraction sign, for example, $7x - (2x + 3)$. A surprisingly high percentage of algebra students interpret this to mean $7x - 2x + 3$ instead of the correct $7x - 2x - 3$. They do better if they recognize that the expression $7x - (2x + 3)$ is shorthand for $7x + (-1)(2x + 3)$.

ASSOCIATIONS   The charge of that nearly ubiquitous particle, the electron, is assigned a value of -1 because Benjamin Franklin in 1750 made an influential guess about static electricity. He subscribed to a "one-fluid" theory to explain the then new discoveries about what appeared to be two different kinds of electricity. Some material objects when charged were observed to repel each other and others attract. Although charges from rubbed amber when transferred to two corks repel, and charges from glass rods also repel, a piece of cork charged by amber produces an attractive force when placed near a piece of cork charged by glass.

In the one-fluid theory, there is only one kind of electric charge; if there is a lot of the fluid in one object and a deficit in the second object, the two objects will attract, but if there is a lot of the fluid in both objects, the objects will repel each other. Because of the symmetry of this theory, it was impossible to tell on the basis of then known information which material produced excess fluid and which experienced a deficit. Possibly because amber's association with electricity goes back to the ancient Greeks (the word *electricity* is from the Greek for "amber"), Franklin guessed that the fluid was removed from the amber and added to the glass; thus, the charge associated with amber became known as negative and the one associated with glass as positive. Even today engineers treat electric current as if it flows from a positive terminal to a negative one.

But Franklin had guessed wrong. After the electron was discovered, near the end of the nineteenth century, it became obvious that electrons rub easily off fur and accumulate on amber, producing an excess of charge. By the one-fluid theory, which is a correct explanation of static electricity, that excess ought to be called positive. But as a result of Franklin's guess, it is called negative. Therefore, the charge of the electron today is called negative.

But the real world is not always as simple as theories about it. A little more than a dozen years after the discovery of the electron it became apparent to Ernest Rutherford and his co-workers that other subatomic particles exist with a charge equal and opposite to that on the electron. They named these particles protons. Although protons do not move easily from one substance to another, their positive charge appears when electrons have been removed, as when a glass rod is rubbed with a silk cloth. Thus, static electricity is caused by the movement of one "fluid" (the electrons), as in Franklin's theory, but there are actually two "fluids" involved—the negative and positive charges.

One of the basic and virtually unassailable conclusions of physics is the CPT (charge–parity–time) theorem, which, like a very few physical statements, is grounded almost entirely in mathematics and not in physics (although it does have a physical basis in the equations of quantum electrodynamics that themselves are highly mathematical). The CPT theorem was put together in bits and pieces around 1950 with contributions from several physicists. One way to describe the CPT theorem is that if you write any equation describing a reaction in particle physics and multiply the charge (C) by -1, a second number called the parity (P) by -1, and the time (T) by -1, the resulting equation will also be a true statement. Another way to say this is that if you assign numbers to C ($+1$, -1, or 0, since quarks do not exist as separate particles), to P ($+1$ or -1, as explained below), and to T ($+1$ for time going forward and -1 for time going backward) to a collection of interacting particles, at the end of the interaction the product of the numbers for C, P, and T will be the same as it was at the beginning. Another way to state this is to say that CPT is conserved.

Parity is often explained as the difference between right and left, but it runs somewhat deeper than that. A slightly better description is that a change in parity would consist of interchanging right and left, up and down, and back and forth. In general, you can say (as with right–left symmetry) that reflection in a mirror changes parity for particle reactions, and that conservation of parity means that the same reactions occur on one side of the looking glass as on the other. The basis of parity is a property that all particles have that is called orbital angular momentum. A right–left symmetrical wave function for a particle implies an orbital angular momentum that is labeled $+1$. If the function interchanges right for left on reflection, then its parity is -1. When two or more particles are interacting as a system, the parity of the group of particles is the product of the individual parities. Conservation of parity when viewed this way means that if a system of several particles has a parity of -1 (or of $+1$) to start with, it will maintain that parity as it changes by particle decay or interactions.

When the CPT theorem was first proved, any physicists who thought about it at all believed that it was true because C, P, and T were all conserved separately, so naturally the combination would be conserved. In 1957, however, in her famous experiment, Chien-Shiung Wu [Chinese-American, 1912–1997] demonstrated that P is not conserved in some interactions. Later in the year a less well known experiment found cases where interchanging all the charges (multiplying every charge by -1 is called "charge conjugation") does not lead to the same reaction as before the charges were conjugated, so C is not conserved by itself either.

This suggested something rather alarming about time. Before the experimental evidence of 1957 the idea that any experiment with a small number of particles should be just as true when carried out backward in time as forward in time had been a bedrock of physics. But if both C and P are not conserved, the implication is that T has an inherent forward or backward direction. Physicists proposed that to save conservation for T, the combination CP must always be conserved, even though C and P are independently not conserved. But seven years later, in 1964, James W. Cronin, [American, b. 1931] and Val L. Fitch [American, b. 1923] discovered a particle interaction that fails to display CP conservation (a discovery that led to their joint award of the Nobel prize in physics in 1980). They found that in a stream that should have the same number of neutral kaons and neutral antikaons—two obscure particles that are not a part of ordinary matter, but that occur in cosmic rays and collisions of particles—there are always slightly more kaons than antikaons, which can happen only if CP conservation fails.

If the CPT theorem is true, which no one really doubts, then time must flow differently from past to future than from future to past, for if CP is not conserved and CPT is conserved, then time must work differently when going forward than when going backward. This may not seem like an odd conclusion to those of us used to living in a steadily forward-moving stream of time, but for particle physics it is completely unexpected. For years after the Cronin–Fitch experiment, the direction of time was only a consequence of a theory, not an experimental observation. Finally, in 1998, a reaction was found with time's arrow sticking out of it. The full implications of this discovery are still being worked out, but it seems likely that the failure of these kinds of symmetry accounts for the observation that there is more matter than antimatter in the universe. (*See also* discussion at **2/3**.)

0   Zero   □

FIELD MARKS   People think of 0 largely in terms of what it is not. It is not a counting number, nor is it a fraction. It is neither positive nor negative. Yet when it comes to theoretical and practical importance, 0 is arguably in the top rank of numbers, and perhaps even the most important.

The idea of 0 as a number is often confused with the closely related concept of nonexistence. Students new to algebra may not recognize that the equation $x + 1 = 1$ has the solution 0, while the equation $x + 1 = x$ has no solution. "No solution" is different from the "solution 0." Here is one way to think of the distinction: your checking account may have a zero balance, but that is different from not having a checking account.

In educational circles the set of numbers that consists of the counting numbers *and* 0 is given its own genus as "whole numbers." But in the higher realms of mathematics this idea is lost. Instead, some mathematicians annex 0 to the natural numbers at least some of the time, while others use 0 only as a species of genus *Integral*. Often the choice of how to classify 0 is a matter of context. Some situations (such as a statement of an infinite sequence or of a general polynomial) call for treating 0 as the first natural number n, while other situations (such as finding the solution to an equation) require 0 to be an integer. Of course, all numbers can also be interpreted as members of higher genera, so if there is an *Integral zero* there is also a *Real zero* (often needed in finding the limit of a function) and a *Complex zero*.

A full-fledged version of 0 was not in place until centuries after the other whole numbers, fractions, decimals, negative numbers, and even such irrational numbers as the square root of 2. There are a few reports of a symbol like "0" used to mean "zero" in Cambodia and Sumatra as early as the seventh century A.D., but most authorities refer to a carved Indian inscription dated firmly at 876 A.D. that employs "0" to mean "zero" as the first authenticated use. Independently, the Maya and other Native Americans began to use a somewhat different symbol for "zero" a couple of centuries earlier, but our present use is a lineal descendant, via the Arabs, of the Indian introduction of 0. A psychologist specializing in the development of the mental concept of number, writing in *Science* in 1998, notes that "The discovery (or invention) of 0, understanding of negative, rational, and real numbers, and development of the calculus took place over thousands of years." Her parenthetical emendation suggests that she views 0 as somehow different from all those other numbers and ideas.

As with other numbers, 0 has somewhat different roles depending on the model, or idea, used to think about numbers:

Measurement 0 (relative). If you think about numbers in terms of measurement, 0 is the starting place, the place from which you begin to measure. This idea is easily seen on an ordinary thermometer. Starting from 0, the temperature may rise to become positive or fall to become negative. Viewed this way, there is nothing special about 0, since if you substitute a Fahrenheit for a Celsius thermometer, the same temperature that is 0°C is 32°F. Similarly, many problem situations use a time of 0, often expressed as $t = 0$, as a starting point. Like the temperature scale, the location of time 0 can be arbitrarily chosen, although it is often taken to be the time at the start of the problem. As with temperature, times can be identified on either side of zero, to give such times as $+2$ (2 units after the start) or -5 (5 units before).

There has been a lot of talk about whether the millennium begins on January 1, 2000, or on January 1, 2001. The argument of the traditionalists is that there was no year 0, so the passage of 2000 years would end on December 31, 2000, with the new millennium beginning the next day. In the Christian calendar the years go straight from 1 B.C. to A.D. 1, with nothing in between. This makes more sense if one recalls that the start of the Christian calendar is attributed to a Syrian monk named Dionysius Exiguus (Dennis the Short) writing in the year he identified as A.D. 525; 0 was unknown at that time in Syria, and so it would not have occurred to anyone to have a year 0. Still, most people look to the 0s as a clue to when the millennium starts. If one goes from 1999 with no 0s to the year 2000 with three of them, then the millennium has started for the same reason that going from 99,999 miles on the odometer to 100,000 is a major milestone in the life of an automobile, or achieving a "first birthday" is in the life of a child. Of course, a car comes from the factory with its odometer set at 0, not at 1, and a child has twelve months of life before the big "1." Thus, this controversy goes round and round. Fortunately, people will tend to forget it for about ninety-nine years after January 1, 2001.

Measurement 0 (absolute). A knowledgeable reader will observe that, for both temperature and time, there is also an absolute 0 that does not permit negatives. Temperature measures the average motion of molecules in a substance, so when all molecular motion ceases, the temperature is as low as it can be, which is 0 on the absolute scale. On the absolute scale, there are no negative temperatures. Similarly, we do not know how to envision time before the big bang that started the universe, so the time of the big bang is also an absolute 0 as far as we know (Buddhists and some physicists may disagree). Similarly, when measuring amounts such as lengths, areas, or volumes, the 0 with which you start is absolute.

Amount 0 (sets). Mathematicians also define numbers on the basis of how many items are present, in which case 0 is the number for a collection that has no items, such as the number of words in English that rhyme with "month." This concept corresponds to an absolute measurement 0. One way to make the correspondence with measurement is to introduce units. In the case of length, for example, if you are measuring a line in centimeters, the length of 0 corresponds to the *amount* 0 of 0 centimeters as well as to the *absolute measurement* 0 with which you start measuring. In set theory, the amount 0 is well defined as the number of elements in the empty set. This definition can become circular, however, unless one proceeds cautiously.

Abstract 0. In addition to measurement 0 and amount 0, the number 0 is also part of an abstract number system, as is true of all numbers. The essence of 0 in most abstract systems is that adding 0 to another number in the system produces no change. This is the defining characteristic of 0 in any system for which an operation of addition exists.

Numeration 0 or *placeholder* 0. Part of the difficulty with 0 seems to be that there are several somewhat different ways to arrive at the concept. From a number point of view, the concepts include the two kinds of measurement 0 that mean "start," the amount 0 that means "none" or even "nothing," and the abstract 0 that is the identity element for addition. But the practice of writing 0 did not enter mathematics or even language from any of those concepts. Instead, it began with a concept that can be called "numeration 0" or "placeholder 0."

Zero emerged in the same way from every system for writing numbers that, like the familiar Hindu-Arabic decimal system, is based on the place of a digit in the written form of the number. The earliest number-notation system that we know of, the cuneiform system of the Sumerians and Babylonians, used this place-value idea; so did an early Chinese system and the system used by the Maya. In these systems, a digit for 1 written in a particular place in the numeral would mean "one," while written in another place the same symbol means "ten," "twenty," or "sixty" (depending on the system); "ten" in the Chinese, "twenty" in the Mayan, and "sixty" in the Babylonian. Similarly, a digit for 2 can mean "two" or "twenty" or "forty" or "a hundred and twenty." This creates a problem: what to do when a place is "empty." At first, both the Babylonians and the Chinese simply failed to deal with this problem, trusting to context: if someone in China wrote that the emperor kept a flock of "539" sacred pheasants using the Chinese digits for 5, 3, and 9, the reader would be expected to know that the flock was really 5039 or 5309 or 5390 pheasants, since the flock of the emperor must be much larger than 539 birds. This unsatisfactory system persisted for centuries in China and the Middle East, but finally both cultures began to use a special symbol to indicate the empty location in a written number. In Babylonian cuneiform this use of the symbol for an empty place began around the time of Alexander the Great (fourth century B.C.), but it started over a thousand years later with Chinese scholars, who before that, however, began simply to leave an empty space. In the interim Chinese system they might have written the number for the emperor's pheasant flock as 5 39 or 53 9. Note that this approach does not work at all for what we would write as 5390. At some time before A.D. 1247 the Chinese scribes are thought to have begun to circle the empty space in a numeral to make it stand out better, which led to the equivalent of 5 ○ 39, 53 ○ 9, and 539 ○. The Indian 0 may have originated from a similar practice. The Maya seem to have used such a special symbol from the beginning, however, although their symbol looks more like an inscribed clamshell than a circled empty space.

This concept of 0 as something for which the only use is to indicate an empty place in a written number persisted until quite recently. One American elementary mathematics series of the 1950s contained the definite, but incorrect, statement "Zero is a placeholder, not a number."

SIMILAR SPECIES   Since both 0 and 1 are identity elements ($a + 0 = a$, $a \times 1 = a$), they are certainly similar. To understand their key difference, it helps to see how they function in a common and important mathematical structure called a group. The idea of a group is important because groups are very simple structures that underlie much of mathematics. Basically a group is a set for which one binary operation (one that combines two entities to get one) on its elements, usually taken to be either addition or multiplication, obeys several simple rules:

The operation on the elements must not lead out of the group.

The operation must be associative: if a, b, and c are elements of the group, with addition taken as the operation, then $a + (b + c) = (a + b) + c$ must be true.

The group must contain an identity element.

Each element a in the group must have an inverse element also in the group (that is, for each a there is a b such that $a + b = 0$).

The identity property and the inverse property for identity are half the properties of the group, and both involve the identity element. But while these identity properties are at the heart of mathematics, they do not necessarily involve the number 0. If the operation is multiplication, the identity element is identified as 1 and not 0. For a multiplicative group the identity property becomes $a \times 1 = a$ and the inverse property is $a \times b = 1$. As a structural element in a group, 0 and 1 have exactly the same properties. Since mathematicians prefer structures to be as empty of meaning as possible—so that they can be applied to situations by assigning them specific meanings—it is possible to resymbolize any isolated group by changing the 1 to a 0 and the operation sign from $\times$ to $+$. Nothing about the group changes when this is done.

For example, a common group is the rotations of a geometric figure, where the operation is interpreted as "followed by." Elements in the group are entities such as "a = one turn of 60° to the left" or "c = four turns of 60° to the right." Then the operation a followed by c could be taken as either $a + c$ or $a \times c$ with no change in meaning.

But in real life and in large chunks of mathematical thought, groups are not quite so isolated. We perceive some group operations as a kind of addition and others as a type of multiplication. Furthermore, when situations occur in which both addition and multiplication are needed, a clear-cut distinction emerges between 0 and 1. While an additive group can and usually does contain an element identifiable as 1, a multiplicative group is never permitted to have an element that could be recognized as 0. For a group to exist, every element must have an inverse with respect to the group operation, but 0 has no inverse with respect to $\times$. There is no corresponding absence for 1, which has the inverse -1 with respect to $+$. Thus, although 0 and 1 are similar species of numbers, they are certainly not the same when you leave the group for larger pastures.

The numbers 0 and 1 are linked in other ways as well. It may seem odd when first learned—and to the thoughtful the fact remains worth contemplating—but for any number a except when $a = 0$, the power $a^0 = 1$. This is often explained as a convenience required to keep the rules of exponents the same for 0 as they are for counting-number exponents, but surely there must be some deeper meaning. That this rule is absolutely required is obvious when you consider the graph of $y = a^x$ for $a \neq 0$. If we did not define a^0 as 1, the graph would not be smooth and continuous as it crosses the y axis

PERSONALITY Addition of 0 to a number makes no change, but multiplication of a number by 0 is totally destructive to the very identity of the number. Despite 0's bland personality for addition and terrorist appearance in multiplication, the uses of the symbol 0 in numeration are perhaps the most practical parts of mental computation, and occur even in standard written algorithms. In this regard, 0 as a symbol can sometimes transcend its importance as a number.

The most common application of 0 in mental computation is in multiplying by a power of 10. In this application 0 is primarily a symbol and a place-holder, even though it is used to accomplish a real operation with numbers. Because multiplication by a power of 10 is so easy to accomplish by annexing 0s to the end of the numeral, this kind of product becomes an important intermediate step in other mental computations, notably in operations with 5 (half of 10) and 10's immediate neighbors, 9 and 11. (These operations are discussed in **Genus *Natural*** at the entries for the respective numbers.)

Multiplication by a power of 10 is also an important step in the common algorithm used for division by a decimal. The basic algorithm for division is a complicated process that is first taught as dividing by a counting number. The division algorithm converts division to subtraction and then subtracts the divisor (the number dividing) from the dividend (the number being divided) by a number of occurrences that is kept as a running tally. When no more divisors can be subtracted, the number tallied is the answer, or quotient. This algorithm with its refrain of "estimate, multiply, subtract, bring down" is virtually impossible to use when the divisor is not a counting number, however. But multiplying both the divisor and the dividend by the same nonzero number will not change the quotient. The arithmetic is easiest when multiplication is by a power of 10. Sometimes, however, a different factor is easier to use. For example, the standard way to divide 168 by 2.5 is to change the problem to

$$25 \overline{)1680}$$

but multiplying by 2 makes the problem $336 \div 5$ and a second doubling makes it $672 \div 10$, so the answer is 67.2 by "moving the decimal point" to divide by 10.

Annexing 0s to the front of a numeral is the inverse operation of annexing them at the tail end, so it is not surprising that annexing 0s to the front divides a number by the power of 10 that has the same number of 0s as that power. But this trick only works as described if the number being divided is between 1 and 10 and does not include 10. Furthermore, you have to be careful to insert the decimal point after the first 0, not in front of it. Thus, dividing 8 by 10,000,000 requires seven 0s, and the answer is 0.0000008. The same idea can be used for any number between 1 and 10, so $3.14159 \div 10^5 = 0.0000314159$. This is particularly useful if you have some reason for converting a number written in scientific notation into ordinary decimal notation, since multiplying by the number expressed as 10 with a negative exponent is the same as dividing by 10 with the exponent replaced by its positive opposite. For example, the number 3.14159×10^{-5}, which is in the proper format for scientific notation, is the same as $3.14159 \div 10^5$, so it becomes 0.0000314159 when it is expressed in ordinary notation.

If a number is greater than 10 or less than 1, you can still divide by a power of 10 by annexing 0s to the beginning—provided that the decimal point is advanced to the left by more places than there are digits in the number being divided. The actual rule is to picture the decimal point as moved the number of places as the number of 0s in the divisor. Thus to divide by 1000, the decimal point is advanced three places, so $27 \div 1000 = 0.027$, while $381 \div 1000 = 0.381$ and $5912 \div 1000 = 5.912$.

ASSOCIATIONS Note that a^0 is not defined when a is 0, and while there are an infinite number of ways that a mathematical expression can be "not defined," most encounters with an expression that is not defined have 0 explicitly or implicitly involved. Perhaps the most common surfacing of this concept is the expression "division by 0 is not defined," which is called to students' attention in class after class from elementary school through college. Thus, 0/0 is also not defined.

In many situations, however, a mathematician might want to assign a definition to 0^0 or 0/0. Indeed, the whole basis of the differential calculus is the common situation in which 0/0 is evaluated and found to be the slope of a curve, which can range from 0 through all possible numbers and even to a vertical slope that is said to be "not defined."

The number 0 has important roles in modern particle physics, since the values 0 charge, 0 spin, and even 0 mass all have significant meanings.

Charge, the force of electromagnetism, is measured in units that can either be integers or multiples of 1/3 (*see* discussions at **−1, 1/3,** and **2/3**). The state of 0 charge is also known as neutral. A particle described as having 0 charge does not respond to the electromagnetic force one way or the other. The neutrino and neutron are well-known neutral particles, each with a charge of 0. It was once thought that a neutron resulted from the union of a −1 charge electron with a +1 charge proton, but scientists now know that this is not the case; instead, a neutron is the unsteady alliance of an up quark with a charge of +2/3 and two down quarks, each with a charge of $-1/3$. On the other hand, when a -1 charge electron encounters a $+1$ charge positron (the antiparticle of the electron), the result is a photon, which has 0 charge. It is not fully clear why some particles respond to electromagnetic forces while others do not, or why charge units are multiples either of 1 or of 1/3.

The quantity known as spin is measured in half-integral units (*see* discussion at **1/2**) and many of the most essential particles, known as bosons, have a spin of 0. Most of the particles that are exchanged with each other to produce forces, all bosons, have spin 0 also, including the pions and kaons. But the gluons that carry the "color" force, which have mass 0, have a spin of 1, as do the heavy particles known as W and Z that carry the electroweak force. At a level of ordinary observation, however, forces are carried by 0-spin photons, 0-spin pions (the strong force that is a manifestation of the color force), and gravitons, which may have 0 spin, but which have not been observed (*see also* discussion of **Genus Natural,** at **4**).

The question of whether or not a neutrino has 0 mass has been one of the most hotly investigated topics in particle physics for several decades. Current experimental results imply a small mass, but the original theory of the neutrino strongly predicts 0 rest mass. The reason is that only particles with 0 mass can travel at the speed of light in a vacuum; furthermore, a particle with 0 rest mass *always* travels at that speed in a vacuum, thus accounting for the absence of right-handed neutrinos (existing left-handed neutrinos would have to stop and turn around to become right-handed). New theories are being built to explain why the neutrino should have a small mass. If the mass of the neutrino is *not* 0 then the curvature of space in the universe *could be* 0, a popular goal of cosmologists, but one they have had a difficult time establishing.

The photon, which is both the particle of light and the carrier of the electromagnetic force, is described in particle physics with a long string of 0s—0 charge, 0 rest mass, and so forth. But of course this does not mean that the photon does not exist.

Genus *Rational*

After the counting numbers, people most often use rational numbers. The positive rational numbers and operations with fractions or decimal fractions representing them are taught in grade school. Despite this, most people use fractions without a clear understanding of what rational numbers are. Even mathematics educators sometimes argue over the exact nature of different representations of rational numbers. People usually encounter rational numbers in various guises— fraction, decimal, percent, ratio—but it is difficult to sort out how these are alike and how different (*see* **Decimal Fractions,** pp. 220–222; **Percent,** pp. 230–233; and **Ratio and Proportion,** pp. 244–246).

A rational number is the quotient of two integers provided the divisor is not 0. We often write such numbers as fractions, in which the names of the two integers are separated by a horizontal bar or a slanted line called a shilling mark:

$$\frac{-2}{3} \qquad \text{or} \qquad -2/3$$

Notice that you cannot tell by looking at the shilling fraction whether $-2/3$ is -2 divided by 3 or the negative of 2/3. This is an example of a serendipitous property of fraction notation that I call "the sloppy rule for fractions." Since division of a negative number by a positive and division of a positive number by a negative both result in a quotient that is negative, it does not matter exactly where you place a negative sign in front of a fraction. The fractions

$$\frac{-2}{3} \qquad \frac{2}{-3} \qquad \text{and} \qquad -\frac{2}{3}$$

all mean exactly the same thing, so you can be as sloppy as you like about where you insert the negative sign.

There is no requirement that these ordered pairs be written in the traditional way with one number, called the numerator, above a line segment and the second number, called the denominator, below it. Sometimes it is revealing to look at rational numbers and operations on them by writing the integers in the other ordered-pair notation, the one used for the pairs of coordinates that locate points in a plane. One reason for this different view is to provide an axiomatic way to explain the rational numbers, as with the ordered pairs used earlier in dealing with the integers (*see* the discussion of integers on pages 191–192 for their definition as ordered pairs). Another reason is to bring a different perspective to operations whose familiarity has made them fade into the mental background. For example, using the coordinate-style ordered pairs, the fraction $-2/3$ would be written either as $(-2, 3)$ or $(2, -3)$. Note that the sloppy rule for fractions from the usual notation no longer works in the coordinate-style ordered pair notation.

One mild surprise when fractions are symbolized in terms of coordinate-style ordered pairs is that the most obvious definition of equality is not adequate for ordered pairs such as (m, n) and (p, q) when such pairs represent rational numbers. We want $(-2, 3)$ and $(2, -3)$ to be equal, although it is clear that $-2 \neq 2$ and also $3 \neq -3$. The correct definition of equality takes care of this: $(m, n) = (p, q)$ if and only if $m = kp$ and $n = kq$ where $k \neq 0$. In the case of $(-2, 3)$ and $(2, -3)$ this definition works with k taken to be -1. It also shows that $(-2, 3) = (-4, 6) = (-8, 12)$ and so forth.

With that behind us, the definition of multiplication is easy, while that of addition is rather more complicated. For multiplication the definition is $(m, n) \times (p, q) = (mp, nq)$, while for addition it is $(m, n) + (p, q) = (mq + np, nq)$. Thus, in an axiomatic treatment, you can begin with a few undefined terms such as *set*, *successor*, *1*, and *0*, and from these develop such concepts as natural number (*see* **Genus Natural**), ordered pair, integer, and rational number without adding additional undefined concepts.

Although we all seem to have a number line for small positive integers wired into the brain, the number line for other numbers appears to be an invention of mathematicians. As was noted in the discussion of negative numbers (*see* **Genus Integral,** 0), many earlier mathematicians thought that the negatives belong somewhere in size above infinity or at least greater than any of the natural numbers. If these mathematicians had had a vision of the modern number line in mind, it would be clear that the negatives begin to the left of 0 in the usual orientation, and not far to the right of the highest counting number. But the rational numbers must also have caused problems for the first people to place them on the number line. The Egyptian notion of fraction, for example, seems not to have anything to do with order of the fractions. When a number such as 4/7 is built from the unit fractions 1/2 and 1/14, the operation is rather more like having a weight of 1/2 unit to which one adds a weight of 1/14 unit to obtain the weight we think of as being 4/7 unit.

It is commonly thought that the fractions first emerged in the trade of items such as olive oil, wool, or metal, although there seems to be no real evidence. The Egyptian idea of unit fractions appears to represent the kind of fractions that might be left over in transactions involving a continuous object (such as olive oil) as opposed to a discrete object (such as a sheep). Thus, one might look at a jar and say that there is 1/5 or 1/3 of the oil left over. The idea of 2/5 instead of the nearby 1/3 or 1/2 seems not to have occurred.

Other trade items, however, may have led to the recognition of the places that the fractions appear on a number line. For example, there is very early archaeological evidence for standard lengths of rope and even for woven cloth. Cloth produced by any kind of loom, even a small hand loom, tends to have a particular width, so it is useful to measure its size by length rather than by weight, volume, or even area. Furthermore, it is easy to show by folding a rope or length of woven cloth that a certain amount beyond a whole number length is 1/4 or 3/4 of that length. Unfolded, the extra length becomes a measurement fraction.

From this concept it is only a short step to the recognition that 2/3 is less than 3/4 and greater than 3/5, although these relationships are hardly obvious when considered in isolation. Even today most people have to translate these fractions mentally to their decimal equivalents of 0.666 . . . , 0.75, and 0.6 to make comparisons. If you stick to common fractions, you can compare these quantities only by rewriting them as equivalent fractions with the common denominator as 60, so the fractions become 40/60, 45/60, and 36/60.

The problem of determining whether two fractions a/b and c/d are equal and, if not, which is greater can be resolved in several different ways. One of the most useful is a general rule that uses replacement of the fractions with integers. This rule is based on a process called "cross multiplication," a term that, like many in mathematics, is used to mean several different operations that have nothing to do with one another (don't tell me that mathematicians are finicky about their definitions and strictly logical!). The reason for the name is that the integers are found by multiplying "cross" the equals sign.

The numbers to be compared are ad and bc. If the two fractions a/b and c/d are equal, then ad and bc are also: $ad = bc$. Furthermore, if the rational numbers are not the same($\neq$), then the inequality sense for the pair ad and bc is the same as for the rational numbers. That is, if $ad > bc$ then $a/b > c/d$ is also true, while if $a/b < c/d$ is true, then it follows that $ad < bc$. Note that for 2/3 and 3/4 the integers to be compared are $2 \times 4 = 8$ and $3 \times 3 = 9$; since $8 < 9$, it follows that $2/3 < 3/4$. Viewed in the other order you could say 3/4 ? 2/3 where the ? represents either of $=$, $>$, or $<$. In that order, cross-multiplying produces $9 > 8$, from which it follows, since $9 > 8$ is true, that $3/4 > 2/3$. The same idea works if one or both of the rational numbers are negative. Consider

$$\frac{-2}{3} \ ? \ \frac{2}{3}$$

Cross-multiplying produces $-6 < +6$ so $-2/3 < 2/3$, just as we suspected.

Once one has a concept of rational numbers on the number line, how to operate with them, and a way of determining which of two rational numbers is greater than the other, a curious property of rational numbers appears that is quite different from the properties of the natural numbers or the integers. Between any two unequal rational numbers there is another rational number. For example, between 2/3 and 3/4 one of the rational numbers is their average: $1/2 \times (2/3 + 3/4)$. Following the rules for adding and dividing rational numbers, it is easy to see that this average is $1/2 \times (8/12 + 9/12) = 1/2 \times (17/12) = 17/24$. Checking with cross multiplication, we find $2/3 < 17/24$ because $48 < 51$ and also $17/24 < 3/4$ because $68 < 72$. Thus, $2/3 < 17/24 < 3/4$. The same operation could be repeated to find a number between 2/3 and 17/24 or between 17/24 and 3/4. From the operations previously discussed, this general rule can easily be established with algebra.

The set of averages is not the only example of a set of rational numbers that can be located between two unequal rational numbers. Another set of rational numbers that has considerable theoretical interest is the fraction whose numerator is the sum of the numerators of two unequal fractions and whose denominator is the sum of the two denominators. In symbols, if the two fractions are *a/b* and *c/d*, the fraction

$$\frac{a + b}{c + d}$$

is a fraction between the two original fractions, and therefore repeating this process will produce an infinite set of fractions. This interesting procedure and some of the fractions that result are discussed further at **1/6** and again at **Kingdom Infinity, $\aleph_0$.**

Commonly Seen Species

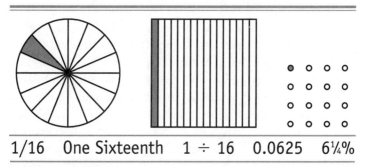

| 1/16 | One Sixteenth | 1 ÷ 16 | 0.0625 | 6¼% |

FIELD MARKS Typically 1/16 is the smallest division on an ordinary foot rule. The reason for this is easy to infer. When you are making a ruler, it is easier to create equal divisions than it is to make marks at 1/3, 1/5, or 1/10 of a small unit, such as an inch. So you divide once to get 1/2. Divide the halves to get 1/4, the 1/4 to get 1/8, and the 1/8 to obtain a division of 1/16. Occasionally one might come across a ruler where one more division has taken place, to produce 1/32, but this is rare as 1/32 inch is too small for ordinary carpentry tools.

SIMILAR SPECIES The number 1/16 is a *unit fraction,* which means that it is formed by dividing a single whole into a number of identical parts and considering one of the parts. All unit fractions have certain traits in common. Each is a "unit" in more ways than just having a 1 as the numerator. Just as one counts with natural numbers by starting with the unit 1 and adding 1 each time to get the sequence 1, 2, 3, 4, 5, . . . for a fraction you can begin with the unit fraction, such as 1/16, and count 1/16, 2/16, 3/16, 4/16, 5/16. . . . This simple sequence is often obscured because of our desire to write all rational numbers in lowest terms (*see* below), so we picture the sequence as 1/16, 1/8, 3/16, 1/4, 5/16, . . .

One of the features of ancient Egyptian mathematics that seems odd to the modern Western mind is the requirement that all fractions except 2/3 (and rarely 3/4) be expressed in terms of sums of unit fractions. The concept of an Egyptian unit fraction, however, was not exactly the same as the modern unit fraction. To form 1/16, the modern mathematician can take any one of sixteen identical parts, but the Egyptian language requires that the part be the first of the sixteen parts. Thus, the concept of 3/16 simply did not occur to the Egyptian mind. Furthermore, Egyptian scribes would never have written 3/16 as an expression we might interpret as 1/16 + 1/16 + 1/16. Instead, the number would have been written with three different unit fractions, since in any series of divisions only one of them might be the sixteenth part. In the case of 3/16, a simple 1/8 + 1/16 would suffice, although, because this means the eighth part followed by the sixteenth part, it is more appropriate to write it as 1/8 1/16 rather than as 1/8 + 1/16. Many fractions written as unit fractions are not so obvious. An example would be 4/7, which becomes 1/2 1/14, with nary a seventh showing. Sometimes these notations can be very sophisticated. One Egyptian way of writing 2/61 was as 1/40 1/244 1/488 1/610, while an old Egyptian problem has as its solution 14 1/4 1/56 1/97 1/194 1/388 1/679 1/776, a number that we would write as $14^{28}\!/\!_{97}$.

Although the Egyptian unit fractions seem very awkward today, their use persisted well into classical times in the mathematics of Greece and Rome.

The other basic measurement fractions are 1/10 in the metric system and 1/12 in the Germanic and English tradition. In cooking one encounters measurements in 1/3 cups as well as in 1/4 cups.

PERSONALITY   It is especially true of unit fractions that they can be seen as a different way of looking at whole numbers. Thus, the same argument for dividing an inch into sixteenths can be given for turning a large sheet of paper into sixteen pages (*see* discussion at **Genus Natural, 16**). When dealing with unit fractions, then, it is easy to transfer an operation to one with whole numbers, while when dealing with whole numbers, one can often switch the ground of battle to unit fractions. (Students sometimes fail to recognize this. If near the end of solving an equation with fractions they reach the step

$$\frac{1}{16}x = 5$$

they will proceed by dividing by 1/16, since they have learned to solve $2x = 8$ by dividing by 2, instead of the equivalent, quicker, and conceptually easier solution of multiplying both sides of the equation by 16.)

ASSOCIATIONS   Although 1/16 as a division used in measurement of length seems to stem from successive halving, 1/16 as the ratio of an ounce to a pound appears to be more of a parallel choice instead of a logical development. The original pound had 12 ounces, not 16.

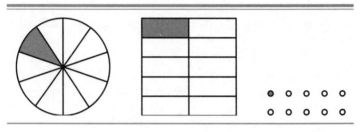

1/10   One Tenth $1 \div 10$ 0.1   10%

FIELD MARKS   The number 1/10, as the multiplicative inverse of 10, is another way to describe the fundamental factor in the decimal place value system and in the international system of measurement (the modern version of the metric system, formally Système Internationale des Unités, hence SI). Just as each place in a Hindu-Arabic numeral has 1/10 the value of the place to its left, so each SI unit of measure is 1/10 of the next larger unit. This has a particular meaning for measurement, since there are only seven base SI units and one often wants to use smaller measures. The prefix for 1/10 is *deci-*. Oddly, however, very few measurements are given in terms of the decimeter, decisecond, deciampere, decikelvin, decimole, or decicandela—the seventh base unit, the kilogram, is born with a prefix, so a tenth of a kilogram is a centigram (10 grams) and not a decikilogram. For most purposes, people are more comfortable with measures that are farther apart than a factor of 10. Thus, the millisecond (1/1000 second) and nanosecond (1/1,000,000,000 second) are used much more often than the decisecond. (My computer spell-checker, I notice, recognizes millisecond and nanosecond as correctly spelled words but has no idea what to do with decisecond.) An exception to the general rule might be made for centimeter (1/100 meter) in relation to the millimeter (1/1000 meter), since a millimeter is 1/10 centimeter and both are commonly used measures—although some writers prefer to write 0.3 centimeter instead of 3 millimeters.

SIMILAR SPECIES   The various negative integral powers of 10, such as $10^{-2} = 1/100$ (which are also natural-number powers of 1/10 since, for example, $(1/10)^2 = 10^{-2} = 1/100$), form a family with familiar properties.

PERSONALITY   Multiplying by 1/10 or by a natural-number power of 1/10 is often described as "moving the decimal point" to the left—one place to the left for 1/10 or as many places as there are 0s in $(1/10)^n$. Thus, $3987.56 \times 1/10 = 398.756$ and $3987.56 \times 1/1000 = 3.98756$, and so forth. Purists point out that one cannot "move the decimal point" in a number, but even purer purists say that there is no reason why one cannot move the decimal point in a numeral. As with annexing 0s to the end of a number to multiply by a power of 10, moving the decimal point to multiply by a power of 1/10 is a case of performing an operation on a number simply by rewriting its numeral. Note also that if the original number is a decimal fraction that is not a natural number (for example, 23.0837), then multiplying by 10 or a power is *not* a matter of annexing 0s but of moving the decimal point, but to the right instead of the left (100 × 23.0837 = 2308.37).

ASSOCIATIONS The tithe, or 10% of income, that good Christians may give their church originated as a tenth part of the harvest, and comes from both Jewish and Roman traditions. Even earlier, the Greeks set aside a tenth part (*dekáte*) as an offering to their gods.

Decimal Fractions

At one time, children were taught a great deal about common fractions (a numerator and a denominator with a line between them), and when their teachers thought they were (or textbooks indicated that they should be) able to add, subtract, multiply, and divide common fractions and mixed numbers (such as 2½), they were deemed ready to tackle decimal fractions. About thirty years ago elementary mathematics textbooks began to bring decimal notation for fractions in much earlier, and many pupils today learn about 0.1 at the same time as they learn 1/10, although they may have studied a few simple fractions with low denominators earlier. For the most part, the only concern in operating with decimals instead of whole numbers is keeping track of the decimal point in the answer. Addition and subtraction require two decimal fractions with the same number of "decimal places," that is, the number of digits following the decimal point. The decimal method produces the same result as does adding common fractions, but it is easier because there is always a common denominator: 43.98 and 30.70, for example, both have the common denominator hundredths, $43^{98}/_{100}$ and $30^{70}/_{100}$.

The rule for multiplication of decimal fractions can similarly be understood in terms of common fractions. The rule is to use the number of decimal places in the product that is the sum of the numbers of decimal places in the factors. For example, the product of 0.23 (2 decimal places) and 0.4 (1 decimal place) is 0.092 (1 + 2 = 3 decimal places). This odd idea seems less strange for those of us brought up on common fractions if we translate the problem to $23/100 \times 4/10 = 92/1000$. Notice that the denominator is 100×10, which can be calculated by counting the 0s in the factors and making sure that there are the same number of 0s in the product. It is not hard to see that this will always happen.

A similar argument shows that the number of decimal places in a quotient is the difference found by subtracting the number of places in the divisor from the number of places in the dividend. Thus, to divide 2.43 by 0.3, the answer would be the same as the answer to $243 \div 3 = 81$ except with 1 decimal place, so $2.43 \div 0.3 = 8.1$. This could become a bit dicey if the divisor has more decimal places than the dividend, so dividing 24.3 by 0.03 would seem to require -1 decimal place, which is hard to do, although a little thought will suggest that 810 is 81 with -1 decimal place. This rule for division of decimals is correct and not all that difficult, but educators found an easier way to handle this issue. First they teach a rule for dividing a decimal by a natural number, which is that the decimal point in the answer is just above the point in the dividend. Of course, this rule only works for division when it is written in the format with a sign like a parenthesis instead of with the sign $\div$. Thus instead of writing $3.51 \div 9$, the student learns to write

$$9\,\overline{)3.51}$$

Then the answer becomes

$$
\begin{array}{r}
0.39 \\
9\,\overline{)3.51} \\
\underline{2\ 7} \\
81 \\
\underline{81}
\end{array}
$$

After students learn how to divide by a natural number, they learn that with decimal fractions you need to multiply the divisor and dividend by an appropriate power of 10 to make the dividend into a natural number. Of course, this is more easily taught as "moving the decimal point" once again, or at least it was before the mathematical purists began to object.

If decimal fractions are so much easier than common fractions, why don't we forget all about common fractions and just stick to the decimals? There are a couple of reasons. Although we eventually become so familiar with using 0.5 for 1/2 that they are about equally intuitive, it is much less clear that, say, 3/8 and 0.375 are the same. In many situations, such as eating a pizza or measuring a piece of wood, the 3/8 says something to you that 0.375 simply does not. But the more significant reason is that decimals for most common fractions can only be expressed as infinite decimals (*see* **Infinite Decimals,** pp. 224–227).

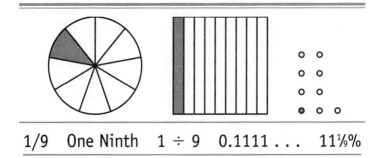

1/9   One Ninth   1 ÷ 9   0.1111 . . .   11⅑%

FIELD MARKS   In the decimal system, 1/9 is the rational number that, other than 0, has perhaps the easiest representation using a decimal point, since it is an infinite string of 1s after the point. As such it is reminiscent of the rep-units among the counting numbers (*see* discussion at **Genus Natural, 11**).

SIMILAR SPECIES   The other ninths are all the other infinite decimals that repeat the same digit, thus, 0.0000 . . . = 0/9; 2/9 = 0.2222 . . . ; 3/9 (= 1/3) is the familiar 0.3333 . . . ; 4/9 = 0.4444 . . . ; 5/9 = 0.5555 . . . ; 6/9 (= 2/3) = 0.6666 . . . ; 7/9 = 0.7777 . . . ; and 8/9 = 0.8888 . . . . Those unfamiliar with infinite decimals may be surprised to see that following this procedure leads to the next one in the series being 9/9 = 0.9999. . . . It is a surprise because we already know that 9/9 = 1 = 1.0000 . . . . Is it possible that 0.9999 . . . and 1.0000 . . . are two ways to show the same number? It is not only possible, but exactly the case.

   In general, when rational numbers are shown as infinite decimals, the concept of a one-to-one correspondence between the number and the infinite-decimal representation fails; for many rational numbers, there are two equally good ways to express the number as an infinite decimal. The numbers for which one-to-one correspondence between infinite-decimal form and fraction (or integer) form fails are those that can also be represented by a terminating decimal (*see* **Infinite Decimals,** below). Rational numbers that can be represented by terminating decimals include all the integers as well as all the rational numbers represented by fractions that have denominators whose only prime factors are 2 and 5 (that is, denominators such as 2, 4, 5, 8, 10, 16, 20, 25, and so forth). For such rational numbers there exist pairs of decimal representations, such as $2 = 2.0000 \ldots = 1.9999 \ldots$ and $1/8 = 0.1250000 \ldots = 0.1249999 \ldots$, as well as $1 = 1.0000 \ldots = 0.9999 \ldots$.

Like 1/4 and 1/25, 1/9 is the reciprocal of a perfect square (*see* discussion at **1/4**).

PERSONALITY Both 0.11 and 11%, although not exact, are sufficiently good approximations for 1/9 to be used in casual computations and are much easier to compute with in most situations.

Infinite Decimals

Mathematics educators and even mathematicians often say that the decimal expansions of 1/16 and 1/10 terminate but 1/9 is an infinite decimal. By this they mean that a terminating decimal can be written with a finite number of digits, while an infinite decimal cannot be exactly expressed, since there will always be another digit needed to get closer to an exact value. From a different point of view, perhaps a more sophisticated one, no decimal expansions of rational numbers ever terminate. Some rational numbers can be expressed as a decimal fraction that continues with an infinity of 1s, such as $1/9 = 0.11111 \ldots$, while others continue with an infinite number of 0s, such as $1/10 = 0.100000 \ldots$. Viewed this way, there is not much difference between decimals that terminate and those that are infinite.

Because 10 has only the prime factors 2 and 5, most fractions cannot be written as a decimal that has a finite number of terms. Whether a decimal terminates or repeats endlessly is determined entirely by the denominator of the corresponding fraction when the fraction is in lowest terms. Consider the natural-number denominators starting with 2. The half terminates because the only factor is 2. But thirds cannot terminate, since 3 is not a factor of 10. Instead, the same digit repeats endlessly: $1/3 = 0.33333 \ldots$ and $2/3 = 0.66666 \ldots$. Next up are the fourths, from 2×2, which terminate after two nonzero terms: $1/4 = 0.25$ and $3/4 = 0.75$. Since 5 is a factor of 10, the fifths terminate after a single term. The sixths have a factor of 3, which makes them nonterminating: $1/6 = 0.166666 \ldots$ and $5/6 = 0.833333 \ldots$. The sevenths, difficult as all numbers involving a 7, repeat a whole cluster of terms over and over (*see* discussion at **1/7**). Since there are three factors of 2 in 8, the eighths terminate after three terms. As noted, $1/9 = 0.11111 \ldots$ and the other ninths also repeat: $2/9 = 0.22222 \ldots$ and so forth.

Skipping ahead, the sixteenths terminate after four terms, the thirty-seconds after five terms, and the sixty-fourths after six. Similarly, the twenty-fifths terminate after two terms and the one-hundred twenty-fifths after three. I trust you see a pattern here.

The common division algorithm shows why the same pattern will repeat over and over. Eventually you encounter a remainder that is the same as in an earlier step. All remainders must be less than the dividend. Therefore, there are a finite number of remainders that must appear in an infinite sequence, so some must repeat.

It is interesting to look for the fractions that repeat a pattern of three digits instead of the single-digit repeat of the thirds and ninths or the double-digit repeat of the elevenths. When you search a few small numbers, you will find a pleasant and, as far as I know, unexplained, surprise. The smallest denominator that produces a repeat of a triple of different digits is $1/27 = 0.037\ 037\ 037\ 037\ \ldots$, and the next smallest is $1/37$, for which the decimal expansion is $0.027\ 027\ 027\ 027$. $\ldots$ This makes it tempting to look at $1/47$, but there seems to be no further surprise of this nature, since $1/47 = 0.0212765957446808510638209787234042531014893617$ $0212765957446808510638209787234042531014893617\ \ldots$ which has 45 digits in the repeating pattern (shown twice here). The repeat begins as soon as a remainder is found that matches an earlier remainder—in this case, the remainder of 6 appears after what seems to the hand-calculator to be an unending sequence: 6, 13, 36, 31, 28, 45, 27, 35, 21, and on and so on. The only hope is that one can be certain that the remainder *will* repeat, because there are only 46 possible remainders that are natural numbers less than 47.

In any case, this example illustrates that eventually all rational numbers produce repeating decimals. The converse, that all decimals that repeat a period of n digits must represent rational numbers, is also true. Rather than attempt a formal proof, let me demonstrate a general method that will always convert a repeating decimal into a rational number, provided one is not too finicky about the infinity part. Let's convert the period-3 repeating decimal $d = 0.123\ 123\ 123\ \ldots$ to a rational number. The trick is to eliminate the repeating part by multiplying by a suitable power of 10 that will give a repeating decimal that begins with a whole number. Then the original decimal can be subtracted from the new product to give a whole number. In this case, we begin by noting that $1000d$ is $123.123\,123\,123\ldots$, so $1000d - d = 999d = 123$, so $d = 123/999$ or $41/333$ in lowest terms. It is easy to see that the same method will work for any repeating decimal.

Of course, if a decimal does not repeat, the method will not work. Since all rational numbers convert to repeating decimals, then any sequence of decimals that does *not* repeat must not be a rational number. Such sequences are easy to construct. Perhaps the simplest follows each 1 with the next counting number of 0s, so that the nonrepeating decimal is 0.10100100010000100000100000010000000100000000001 and so forth. Another one that has been suggested and has even been studied by mathematicians is the sequence of counting numbers written as a decimal, beginning 0.1234567891011121314151617181920212223242425 and continuing without stopping. At this point, the main thing to note is that such numbers cannot be rational. More will be said later (*see* **Genus Real**).

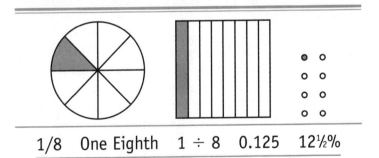

1/8   One Eighth   1 ÷ 8   0.125   12½%

FIELD MARKS Because 8 is the cube of 2, three dissections of a symmetrical and finite portion of space such as a cube or sphere into two parts each—think of left and right, front and back, top and bottom—produces parts equal to eighths. For example, the left–front–top part of a cube is an eighth of a cube. This has implications for the way we organize our division of three-dimensional space, which is into eighths—the three mutually perpendicular planes that are used to separate "3-space" create regions called *octants*, each of which (though infinite) may be thought of as 1/8 of all space. In this view, all space is pictured as a symmetrical analog of a cube or sphere (of course, we do not actually know this to be true, but hardly anyone ever considers alternatives to a symmetrical universe).

Similarly, and perhaps more surprisingly, any finite three-dimensional figure, whatever its shape, can be separated into eight equal-volume parts by three mutually perpendicular planes. Symmetry is not required. This is a version for 3-space of the ham-sandwich theorem, which states that no matter how sloppily the pieces of bread and the slice of ham on a sandwich are arranged, there is a way to cut the sandwich with a knife that simultaneously divides both pieces of bread and also the ham into regions of equal areas. The bread and the ham do not even have to be rectangular, but may be any shape. (*See also* discussions and related proofs at **1/4** and **1/2**.)

SIMILAR SPECIES The other eighths are 2/8, 3/8, 4/8, 5/8, 6/8, and 7/8, but three of these are usually seen as fourths or halves (2/8, 4/8, and 6/8). Thus, the similar species for the eighths are the fourths and the halves.

All fractions reduce to lower denominators unless the numerator and denominator have no common factors, other than 1, which is always a common factor and thus trivial and ignored. Recall that numbers with no common factors, such as 9 (3 × 3) and 22 (2 × 11), are called relatively prime. When you consider fractions between 0 and 1, the number of relatively prime pairs increases irregularly. For each *prime* denominator considered, the number of reducible fractions in that interval is always 1 less than the prime. Thus, for the nine denominators 2, 3, 4, 5, 6, 7, 8, 9, and 10 the numbers of irreducible fractions between 0 and 1 is, in the same order, 1 (1/2), 2 (1/3, 2/3), 2 (1/4, 3/4), 4 (1/5, 2/5, 3/5, 4/5), 2 (1/6, 5/6), 6 (1/7, 2/7, 3/7, 4/7, 5/7, 6/7), 4 (1/8, 3/8, 5/8, 7/8), 5 (1/9, 2/9, 4/9, 5/9, 7/9, 8/9), and 4 (1/10, 3/10, 7/10, 9/10).

The other fractions whose denominators are powers of 2, such as 1/2, 1/4, 1/16, 1/32, and so forth, are also used extensively in measurement, in part because of the ease of dividing a segment by folding it over, producing lengths that have successive powers of 2 as the denominators.

PERSONALITY It is extremely convenient to know the decimal or percent equivalents to all four of the irreducible eighths between 0 and 1. These are (in percent form) 1/8 = 12½%; 3/8 = 37½%; 5/8 = 62½%; and 7/8 = 87½%. Not only are these frequently encountered themselves, but also they are useful in making many good approximations. The percents 35%, 65%, and 85% are near to 3/8, 5/8, and 7/8, which are often easy to compute mentally, but actually 35% = 7/20, 65% = 13/20, and 85% = 17/20, all difficult because of the primes in their numerators. Thus, if you wanted to estimate 65% of 286, you could think that it is somewhat greater than 5/8 of 286. You can approximate 5/8 of 286 by thinking of 5/8 as 1/2 + (1/4 of 1/2). Thus, to find 5/8 of 286, take half of 286, which is 143, and add 1/4 of 143, which is slightly less than 36 (36 × 4 = 144) to get 179 as the approximation. Thus, 65% of 286 is somewhat greater than 179. This may seem complicated but in practice is faster mentally than multiplying 286 by 0.65, the traditional way to find the exact answer (286 × 0.65 = 185.9).

ASSOCIATIONS For many Americans the strongest association with 1/8 is a slice of pizza. My local pizzeria uses four cuts to slice a large pizza into eight slices (each 1/8 of the pizza) and three cuts for a small pizza, which makes a slice 1/6 of the pizza. Some pizza cutters, however, use eight slices to make sixteen skinny pieces, and therefore two slices are required to make 1/8 pizza.

Sometimes 1/8 is the smallest division on an inferior foot rule. Nearly all carpentry work is measured to eighths, but it is helpful to have the sixteenths as well because they make it easier to find the nearest eighth.

Percent

Some years ago an editor asked me to add percents to a table of common fractions and their decimal equivalents. "But they are exactly the same," I protested, since to me the line in the table 1/5 = 0.2 is the same as saying that 1/5 = 20%. I now realize that he was correct, since people perceive percents as substantially different from decimal fractions. Perhaps 0.125 and 12½% seem interchangeable, but the percent has that common fraction 1/2, which would never appear in a decimal. For 33⅓%, the decimal equivalent is a repeating infinite decimal, intellectually much more challenging. Finally, although a decimal fraction represents a number complete in itself, a percent does not stand for a number in the ordinary sense, another reason some folks like percents better.

"Percent" is a relationship, a comparison of one number to another as if the second number were 100; a "percentage" is the amount that is determined by this comparison. Thus, "40 percent [I'm spelling this out for clarity] of the estate is in securities" describes the portion of the estate for which there are stock certificates; if the estate is $200,000, the sum of $80,000 is the amount, the percentage, of the estate that's in blue chips.

This is a nice distinction, and one important to bear in mind—which may be tricky because conventional English usage (supported by such entities as the U.S. Bureau of the Census) uses the word "percent" only when preceded by a number and "percentage" only when not. The problem with the conventional English rule is that it is designed for people who are writing about amounts and not for people who are writing about the concepts of mathematics. Only those whose task it is to explain the meaning of the concepts percent or percentage need to use the word "percent" without a numeral preceding it. The word "percentage" never has a numeral preceding it because a percentage is a number, but a percent, in the abstract, is not. Thus, for teachers and others who speak or write about mathematics, the question is not whether or not a numeral precedes the term, but what the term means. If it is a ratio, it is a percent, so 25% (the ratio of 25 to 100) is a percent. But 25 percent of 100 voters is the percentage 25. "What percentage of the voters supported teaching the concepts of mathematics instead of simply how to compute? Sadly, only 25 percent of them preferred concepts."

The relations between a percent and a number may also seem mysterious. Common advice for operations involving a percent sign is to convert to a decimal fraction and multiply. Some may see no point in replacing 25% with 0.25, but the logic is there: since 25% is not actually a number, you cannot multiply by it directly.

Notice that percents do not add or subtract in any meaningful way. If you find a store that advertises "Everything 10% off—even sale items" and you find a $10 belt that is already on sale at 20% off, do not expect to get the belt for 20% + 10% = 30% off. The correct calculation is $10 − ($10 × 0.2) = $8 for the original sale price and $8 − ($8 × 0.1) for the selling price of $7.20.

You can solve such a problem more generally if you replace 20% off with 80% of the price and 10% off with 90% of the price. After changing to decimals, the problem becomes 0.8 × 0.9 × $10 = $7.20. This method shows that 20% off combined with 10% off is always the same as 72% of the sale price, or the same as 28% off. Still, you would not say that 20% + 10% = 28%, although as method of calculation that oddball sum would be correct.

A mathematician I know once made careful distinctions among a fraction that represents a number, such as 1/8; a rate, such as walking 1 mile in 8 minutes; and the ratio 1:8 that might compare the number of cashews to peanuts in a box of nuts (1 cashew for each 8 peanuts). These distinctions may be correct but are almost impossible to keep a firm grip on. Still, it is sensible to say that percent is a kind of ratio and not a rate or a fraction. Typically a rate deals with two types of amount that are different entities, such as miles and minutes; but percent considers a whole that consists of one kind of thing (which may have subsets, as "voters over age fifty" and "first-time voters" are part of the set "voters").

If there are two different entities completely, such as two-by-fours and bricks, percent is not a useful concept. Rate can still be useful, for in building a brick house you might know that you will need 200 bricks for each two-by-four, for example, which is a rate; but the question "What percentage of the two-by-fours are bricks?" is clearly nonsense, whereas you might sensibly ask "What percentage of the bricks are to be broken in half?" or "What percentage of the two-by-fours should be 12-feet long?"

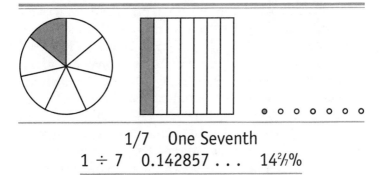

1/7   One Seventh
1 ÷ 7   0.142857 . . .   14²⁄₇%

FIELD MARKS As you go through the rational numbers expressed as common fractions, starting with 1/1 and continuing in the order of slowly increasing denominators—1/2, 1/3, 2/3, 1/4, 3/4, 1/5, 2/5, 3/5, 4/5, 1/6, 5/6—you can easily convert the common fractions to decimal fractions. As decimals, these familiar common fractions all repeat either 0, 3, or 6 endlessly (decimals that repeat 0 are said to terminate). Then comes 1/7. If you are unfamiliar with the conversion and simply begin to calculate the decimal by dividing 7 into 1, you will be alarmed to see no apparent repetition at all. Not until you have reached seven places is there a repeat of even a single digit: 0.1428571. But after a few more divisions, the pattern does become apparent: 0.142857 142857 142857 . . . .

It is easy to show that the greatest possible number of digits in a repeating pattern for a rational number expressed as a decimal when the denominator is n is $n - 1$. The common division algorithm when both the divisor and dividend are positive integers involves the largest multiple of the divisor that is less than the number "brought down" with the annexed 0. Since each partial dividend consists of a remainder from the previous subtraction, there can be no more than 1 less than n remainders.

SIMILAR SPECIES Since $7 = n$ is prime and not a factor of 10, there exist $n - 1 = 6$ proper fractions that have 7 as a denominator: 1/7, 2/7, 3/7, 4/7, 5/7, and 6/7. These all convert to decimals with a repeating denominator in which the period of repetition is six digits long and the digits are the same as those for 1/7 and in the same order (although beginning at different places: $2/7 = 0.2857142 \ldots$; $3/7 = 0.4285714 \ldots$; and so forth. This is a direct consequence of their origin, and the reason can easily be seen from the ordinary algorithm used for long division when the division does not "come out even." The combination of bringing down more 0s and the limited number of remainders less than 7, which can be only 1, 2, 3, 4, 5, and 6, mean that trial divisions must eventually be one of the following: $10 \div 7 = 1$ r 3, $20 \div 7 = 2$ r 6, $30 \div 7 = 4$ r 2; $40 \div 7 = 5$ r 5; $50 \div 7 = 7$ r 1, and $60 \div 7 = 8$ r 4. So the remainders, in this order, are 3, 6, 2, 5, 1, 4. Whenever one of these remainders repeats in the algorithm, the sequence starts again (*see also* discussion at **1/9**).

As you continue through the prime denominators, others appear in which the number of digits in the repeating period is the theoretical maximum of $n - 1$ where n is the denominator. These fractions with the maximum periods begin 1/17, 1/19, 1/23, 1/29, 1/47, 1/59, 1/61, ...

PERSONALITY Although nearly all computations that involve 7 in any way are more difficult than those with no 7s, it is sometimes possible to substitute another number that is nearby to make a simple approximation. Both 1/8 = 0.125 and 1/6 = 0.16666 . . . are neighbors, but not very close ones, to 1/7 = 0.142857 . . . . But note that the average of the two is 0.14583333 . . . , only slightly greater than 1/7. If a number is a multiple of both 6 and 8, such as 192, then averaging 1/6 and 1/8 is fairly easy (1/6 of 192 = 32 and 1/8 of 192 = 24, so 1/7 is about 28). But even in this case it may be easier to find 1/7 of 192 directly, which would give 27 r 3 or 27³⁄₇

ASSOCIATIONS The old idea that the seventh son of a seventh son has unusual powers has nothing to do with 1/7: in English the same words "one seventh" are used to mean both the ordinal seventh, as here, and the fractional seventh (which could relate to a family only in the statistics beloved of sociologists seeking an average). Although there is no fractional relative of "first" and the ordinal "second" is clearly different from the fraction "half," after "two" and "second" the same words are used for ordinals as for fractions: *third, fourth, fifth,* and so forth.

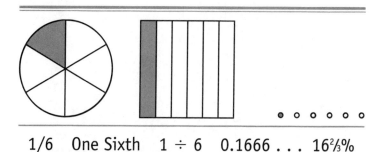

1/6   One Sixth   1 ÷ 6   0.1666 . . .   16²⁄₃%

FIELD MARKS The number 1/6 is often called the first Bernoulli number. As such, it should logically be labeled B_1, but for reasons that will become clear, 1/6 is more commonly known as B_2. This is based on a common notation that would make the *first* Bernoulli number 1 but labels the second, which is 1/2, as B_1. (In B_1 or in any situation where a subscript such as $_1$ is used to specify a variable or a constant, the subscript is called an index.) As I explain what the Bernoulli numbers are, this small confusion in notation *may* begin to make some sense.

First, there is a strong tendency in mathematics to name entities after the wrong mathematician. Bernoulli numbers were not introduced by any of the many mathematicians and physicists of the famous Bernoulli family, although Jacques Bernoulli [Swiss, 1654–1705; also known as James or Jakob] is the occasion for the name, since he described them in a widely read work. Bernoulli got the numbers from another Swiss mathematician, Johann Faulhaber, and properly credits Faulhaber for the development. Faulhaber originally developed these numbers as coefficients in a general equation that gives the sum of the first n powers of the natural numbers. The formula is a generalization of such rules as $1 + 2 + 3 + \ldots + n = 1/2\ (n^2 + n)$; $1^2 + 2^2 + 3^2 + \ldots + n^2 = 1/6\ (2n^3 + 3n^2 + n)$, and $1^3 + 2^3 + 3^3 \ldots + n^3 = 1/4\ (n^4 + 2n^3 + n^2)$ to any power of n. Faulhaber's formula can be written as follows:

$$1^{k-1} + 2^{k-1} + \cdots + n^{k-1} =$$

$$\frac{1}{k}\, n^k \times \mathbf{1} + \frac{k!}{(k-1)!} n^{k-1} \times \frac{\mathbf{1}}{\mathbf{2}} + \frac{k!}{2!(k-2)!} n^{k-2} \times \frac{\mathbf{1}}{\mathbf{6}} +$$

$$\frac{k!}{3!(k-3)!} n^{k-3} \times \mathbf{0} + \frac{k!}{4!(k-4)!} n^{k-4} \times \frac{\mathbf{-1}}{\mathbf{30}} + \cdots$$

The first few Bernoulli numbers are in boldface. Note the resemblance of this formula (except for the Bernoulli numbers) to the binomial formula. The numbers of the form

$$\frac{k!}{r!(k-r)!}$$

are the binomial coefficients.

The easiest way to compute the Bernoulli numbers appears to be a recursion method that is in fact based on a version of the binomial formula. You need to assume that $B_1 = 1/2$ for this method to work, which is no doubt why the first Bernoulli number of 1 is ignored. Then if k is a natural number greater than 1, you can write formulas that are identical to the expansion of $(B-1)^k$ but replacing B^k and the various other powers such as B^{k-1} with the corresponding Bernoulli numbers B_k, B_{k-1}, and so forth. Set this new formula equal to B_k and plug in the previously computed values, which leaves you with what you need to find B_{k-1}. Thus, to find B_2 you think of the binomial expansion of $(B-1)^3$, which is

$$B^3 - 3B^2 + 3B - 1$$

and use this to write the formula

$$B_3 - 3B_2 + 3B_1 - 1 = B_3$$

Notice that the B_3 term drops out. If you have already agreed that $B_1 = 1/2$, then this formula becomes $-3B_2 + 3/2 - 1 = 0$, which when solved gives $B_2 = 1/6$. Then to find B_3 think of the expansion of $(B-1)^4$, or

$$B^4 - 4B^3 + 6B^2 - 4B + 1$$

from which you derive the formula

$$B_4 - 4B_3 + 6B_2 - 4B_1 + 1 = B_4$$

From the previous formula you know that $B_2 = 1/6$ and B, so the formula becomes

$$-4B_3 + 1 - 2 + 1 = 0, \text{ so } B_3 = 0.$$

SIMILAR SPECIES In addition to 1 and 1/2, which are often ignored as Bernoulli numbers, and 1/6, the other Bernoulli numbers are 0 (which appears as the Bernoulli number for every B_{2n+1}, that is, for each Bernoulli number with an odd index greater than 1) and $-1/30$, 1/42, 1/30, 5/66, 691/2730, 7/6, 3617/510, and so forth.

PERSONALITY Computations with 1/6 are not easily handled by converting to decimals—1/6 is closer to the awkward prime 0.17 than to 0.16 or 0.15, which are fairly easy to use. A factor of 1/6 can, however be treated as $1/2 \times 1/3$. For example, to find 1/6 of 78 you might begin by thinking 1/2 of 78 is 39 and 1/3 of that is 13.

ASSOCIATIONS Book designers and compositors measure art, layout, and type in picas and points instead of fractions of inches or decimal parts of a meter. A pica, which is divided into 12 points, is a little short of 1/6 inch. If you look closely at a pica rule, you will see that a length of 6 picas is quite close to an inch, while 60 picas is about 1/16 inch short of 10 inches. For practical purposes, the relation of 6 picas (or 72 points) to an inch is sufficient, since there's no need to make the conversion, which would only make the measurement less accurate. As a matter of interest, 1 pica equals 519/31,250 inch exactly.

The common die used in many games has six faces, so the probability of any one of the faces turning up is 1/6. But in many dice games, two dice are used and the sum of the dots on the faces is the important number. One reason that 7 is considered lucky is that the probability of throwing a 7 is also 1/6, the highest probability of any combination of two dice.

Farey Sequences

This seems about as good a place as any to discuss briefly a way of thinking about positive rational numbers that has been productive in a variety of ways. List all the lowest-term fractions from 0 through 1 that have a denominator less than or equal to a particular natural number. The result is a short sequence of rational numbers that is different for each natural number. The sixth such sequence is a typical example. Here the maximum denominator is 6, so the list includes all the fractions in lowest terms that have denominators of 6, 5, 4, 3, 2, or 1:

$$\frac{0}{1}, \frac{1}{6}, \frac{1}{5}, \frac{1}{4}, \frac{1}{3}, \frac{2}{5}, \frac{1}{2}, \frac{3}{5}, \frac{2}{3}, \frac{3}{4}, \frac{4}{5}, \frac{5}{6}, \frac{1}{1}$$

Such a group of numbers is known as a Farey sequence after a British geologist interested in mathematics named John Farey, who, in 1816, first described such sequences.

Farey also published a statement (but not the proof) of the most striking property of the Farey sequences, which is that the middle term of any three terms of any Farey sequence is always equal to a fraction with a numerator that is the sum of the numerators of the other two terms and a denominator that is the sum of the denominators of the other two—in symbols, if a/b, c/d, e/f is the subsequence of three consecutive terms from any Farey sequence, then $c/d = (a + e)/(b + f)$. At random from the sixth sequence, for example, take 3/5, 2/3,

3/4. Farey's theorem, proved by the French mathematician Augustin-Louis Cauchy [1789–1857] soon after Farey stated it states that $2/3 = (3 + 3)/(5 + 4) = 6/9$, which is indeed true. Here are the first five of the Farey sequences:

$$\frac{0}{1}, \frac{1}{1}$$

$$\frac{0}{1}, \frac{1}{2}, \frac{1}{1}$$

$$\frac{0}{1}, \frac{1}{3}, \frac{1}{2}, \frac{2}{3}, \frac{1}{1}$$

$$\frac{0}{1}, \frac{1}{4}, \frac{1}{3}, \frac{1}{2}, \frac{2}{3}, \frac{3}{4}, \frac{1}{1}$$

$$\frac{0}{1}, \frac{1}{5}, \frac{1}{4}, \frac{1}{3}, \frac{2}{5}, \frac{1}{2}, \frac{3}{5}, \frac{2}{3}, \frac{3}{4}, \frac{4}{5}, \frac{1}{1}$$

You might test a few subsequences of three consecutive terms to convince yourself that Farey's theorem is true.

Among other properties of the sequences that you might test is that for any two consecutive terms from one of the series the cross products are consecutive integers—that is, if the two terms are a/b and c/d then ad and bc are consecutive. Also, note that the "symmetric pairs" of rational numbers, with 1/2 as the "line of symmetry," all add to 1; hence, except for the first Farey sequence, all Farey *series* such as $0/1 + 1/3 + 1/2 + 2/3 + 1/1$ sum to a number that is 1/2 more than a natural number.

(A sequence and a series are two related ideas, sometimes confused by people unfamiliar with them. A sequence consists of one term after another, perhaps infinitely, usually written with commas separating terms. When we indicate the natural numbers with 1, 2, 3, . . . , it shows the sequence of natural numbers. A series is just the sequence with plus signs instead of commas, also known as an indicated sum. Thus the series of natural numbers is $1 + 2 + 3 + \ldots$, which when extended without bound has no actual sum; although if you stop extending at any point, the sum is the triangular number for where you stop (if you stop at 3 the sum $1 + 2 + 3$ is 6, also $3^\triangle$). Sometimes readers confuse the series with its sum, however. Continuing the series $1 + 1/2 + 1/4 + 1/8 + \ldots$ without bound produces the sum 2, but 2 is, of course, not the series, which is just the *indicated* sum.)

You can use the Farey theorem to obtain terms that will be added to form the next sequence from the terms of a given sequence, but you have to throw out terms that don't work. Let's start with the sixth sequence, which we visited first:

$$\frac{0}{1}, \frac{1}{6}, \frac{1}{5}, \frac{1}{4}, \frac{1}{3}, \frac{2}{5}, \frac{1}{2}, \frac{3}{5}, \frac{2}{3}, \frac{3}{4}, \frac{4}{5}, \frac{5}{6}, \frac{1}{1}$$

To form the seventh sequence from this one, for example, use the Farey theorem to form new terms. By the theorem the first two terms of this sequence, 0/1 and 1/6, have a term between them, which is $(0 + 1)/(1 + 6) = 1/7$. The same technique used on the second and third terms of the sequence, 1/6 and 1/5, produces $(1 + 1)/(6 + 5) = 2/11$, which is not a part of the seventh Farey sequence because its denominator is too large—so throw it out. The third and fourth terms, 1/5 and 1/4, don't work either, since they produce a term of 2/9, but the fourth and fifth terms, 1/4 and 1/3, are productive, since 2/7 is a member of the sequence. In general, if you just use the pairs whose denominators sum to 7 you will get all the new terms that need to be added to the sixth series to obtain the seventh.

Several properties of Farey sequences that are also useful are easily observed from the examples just given. One is that the order of fractions in a Farey sequence is specific for any particular sequence, since the fractions represent different numbers and they are arranged in order of size. A second is that there are as many Farey sequences as there are natural numbers. Also, every rational number from 0 through 1 is a member of some Farey sequence. Finally, if you consider the nth Farey sequence as points on a number line, then the points that are added to form the $(n + 1)$th sequence are in between the points previously chosen, showing in another way that there is always another point between two rational numbers. (For an interesting extension of the idea of Farey sequences, *see* the discussion at **Kingdom Infinity, $\aleph_0$.**)

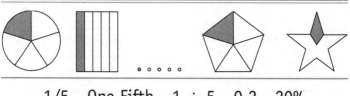

1/5   One Fifth   1 ÷ 5   0.2   20%

FIELD MARKS Think 20%. The close relationship between 5 and 2 that exists throughout a decimal system of numeration reappears in the fractions in a rather symmetric way. Not only is 1/2 the same as 0.5, but 1/5 is also the same as 0.2.

The decimal equivalent also helps keep track of where 1/5 is on the number line or on a ruler or when measuring ingredients in a recipe. The equivalence of 1/4 to 0.25 shows that 1/5 = 0.2 is just a bit smaller than 1/4. It is not as small as 3/16, since 3/16 = 0.1875, but 3/16 is the nearest neighbor to 1/5 on the average ruler or measuring cup.

PERSONALITY In computation, circumstances determine whether it will be better to divide by 5 or multiply by 0.2.

The fifths are also conveniently related to 60. Thus, a fifth of an hour, or twelve minutes, is a whole number of minutes. Similarly, a fifth of a straight angle, or 180°, is 36°, another convenient counting number. In radian measure, $\pi/5$ is exactly 36°.

ASSOCIATIONS Pythagoras discovered that music and harmony are directly related to the counting numbers or, as we might say today, to the rational numbers. He observed that when strings of the same tension and thickness are in a ratio of 2 to 3 (or, to express this relation as a rational number, the shorter is 2/3 the longer), the interval is harmonious. Today we call that interval a fifth, which has nothing to do with the fraction fifth, but is the ordinal number that indicates the position of the note that comes on 5 when you begin counting with 1. Ordinal begins first, second, third, fourth, fifth, telling the order of entities. As noted for the seventh son of a seventh son (*see* discussion at **1/7**), there is only a linguistic connection between the ordinal fifth and the fraction 1/5 that is also called a fifth.

Back to the music. Probably Pythagoras worked with a string under tension. With such a string, if you "stop" it 1/3 of the way along the string, as a guitarist would push it against the fret, and pluck in turn the longer and shorter portions (which are in a ratio of 2 to 1), you produce an octave interval. Vibration at the octave is double the frequency of that at the lower note and sounds to our ears as the same note but higher. If one holds the string 2/5 or 3/5 of the way the ratio of the shorter part of the string to the longer is 2 to 3, producing the interval called a fifth (on a piano it is the interval from C to the G above it or from F to the C above it, for example). This also sounds harmonious and is an interval often used in melodies.

All the notes of the scale can be produced by placing the hold so that the ratios are natural numbers. From this beginning, the Pythagoreans reached the conclusion that the entire universe is based upon natural numbers and their ratios. This concept was soon totally undermined, ironically by a discovery of the Pythagoreans themselves (*see* **Irrational Family,** pp. 274–278).

People who remember the days before metrification probably think of whiskey or other liquor when they hear the phrase "a fifth." The mildly curious may have wondered about the name, since a fifth—a bottle nearly as large as the liter bottle commonly sold today—is clearly not a fifth of a quart of booze. Instead, a fifth of whiskey is fourth fifths of a quart, and hence it is one fifth of a gallon (there are 4 quarts in a gallon, 1/5 of 1 gallon = 1/5 of 4 quarts = 4/5 of 1 quart).

Ratio and Proportion

Today most mathematicians use the concepts of ratio and common fraction to mean the same thing and tend to rely on operations with fractions to handle problems about ratios. But before common-fraction notation became popular, ratios were viewed as fundamental and often had their own rules and algorithms, such as one called the "rule of three" that was endemic in the Middle Ages (it seems likely to have been developed by the Indian mathematicians). Even today mathematicians refer to *rational numbers* and not to *fractional numbers*.

The rule of three is actually a statement about what we call a proportion. Technically, a proportion consists of a statement that two ratios are equal. Today it would be represented by equal fractions. In the rule-of-three days, it would be stated as a relationship given among three terms. The first and third terms had to be of the same kind (such as money or some measurement). In that case, one can solve the problem by multiplying the second and third terms and dividing by the first to get the desired answer. Expressed as fractions, this rule states that if $a/b = c/d$, where a and c are the first and third terms, then d is equal to bc/a. This rule was taught for centuries to merchants with no explanation why it worked. Indeed, without fraction notation and the rules for fractions, it is not so easy to understand.

The notation for ratio and proportion that was popular more recently—and that is still seen outside mathematical circles—does not help much. The ratio of a to b is shown as $a : b$ and the proportion $a : b :: c : d$. If you did not know that this same idea could be expressed with common fractions, it would be necessary to know special rules to handle these problems.

The requirement for the rule of three that the first and third terms be of the same kind is peculiar to ratio and does not apply to fractions. It is necessary so that the rule of three will produce an answer with sensible units. A distinction is sometimes made between fractions and ratio along these lines: a fraction is a part of one thing, but a ratio is a comparison of two things. Thus, one can have a 2/3 of a jar of olive oil (a fraction) but 2 jars for 3 dinars (a ratio). If you are dealing with a proportion, the problem to be solved is usually of the form "If you can buy 2 jars of oil for 3 dinars, how many jars can you buy for 15 dinars?" Stated this way, the problem is of the form $2 : 3 :: ? : 15$. To use the rule of three, however, the merchant was taught to look for the terms with the same denomination, in this case 3 dinars and 15 dinars, and make them the first and third terms, so the problem is thought of as $3 : 2 :: 15 : ?$, which has as its solution $(2 \times 15) \div 3 = 10$ jars.

Other language that was used for proportions included the terms *extremes* for the first and last terms and *means* for the two middle terms, so an ancient Greek problem known as "division in the extreme and mean ratios" meant to divide a line ACB so that the proportion AB : AC :: AC : CB is true (*see* discussion at **Genus Real, 1.61803 . . .**). Similarly, the cross-product rule for fractions is the same as the rule that the product of the means equals the product of the extremes.

Although modern fraction notation greatly simplified the operations of ratio and proportion, since a large number of special rules can be replaced with the algorithms for operations with fractions, the psychological idea of ratio and proportion is very powerful. I have learned that I am not the only one who survived most of seventh- and eighth-grade mathematics by turning every problem into ratio and proportion. And there is nothing wrong with doing this, especially if you also can turn the problem back into fractions to find the solution. It is surprising how many problems can be solved by thinking "A is to B as C is to D," which is the essential proportion.

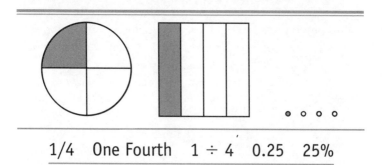

1/4   One Fourth   1 ÷ 4   0.25   25%

FIELD MARKS The number 1/4 is reciprocal of the smallest perfect square after 1. A *reciprocal*, or multiplicative inverse, is the number formed by dividing a given number into 1. The mathematical reciprocal shares with other meanings of the word "reciprocal" the idea of interchanging, just as the expression "each other" is known as a reciprocal pronoun. If you think of 4/1 as the rational number equal to 4, then the interchanging of 1/4 and 4/1 shows that each is the reciprocal of the other: that is, each is formed by interchanging numerator and denominator.

There is another notation that is useful for reciprocals, which is the exponent -1. When used with a number, such as 4, this exponent indicates the reciprocal, so $4^{-1} = 1/4$. It is necessary to specify that this is the meaning of the exponent -1 "when used with a number" because this exponent has a different meaning in some other contexts, such as a trigonometric function. But for any number x, rational or otherwise, $x^{-1} = 1/x$.

The exponential notation -1 is a specific example of the general notion of rational exponents. The number 1/4 can be expressed in many different ways using rational exponents, since negative exponents create various powers of reciprocals while fractional exponents can be used to indicate roots such as square roots, cube roots, and beyond. For example, $1/4 = 4^{-1} = 2^{-2} = (1/16)^{1/2} = 16^{-1/2} = 64^{-1/3}$.

SIMILAR SPECIES The reciprocals of the other perfect squares—that is, 1/9, 1/16, 1/25, and so forth—have similar characteristics.

PERSONALITY People tend to be aware that 1/4 is the same as 25% for most computational purposes, taking 1/4 of an amount to find 25% of it instead of multiplying by 0.25. The other side of this relationship is sometimes also useful. If one wants to find 4% of an amount, it may be easier (depending on the actual amount) to figure 1/25 instead of multiplying by 0.04. An example would be finding 4% of 125 or 150 or 625, which can easily be seen to be 5, 6, or 25 by thinking how many 25s there are in each of the numbers. For example, since there are five 25s in 125, then 4% of 125 is 5.

ASSOCIATIONS It is always possible to divide any finite plane region into fourths using two perpendicular lines (*see also* discussions at **1/8** and **1/2**). This can be done as a consequence of a fundamental idea that, in its commonest form as a theorem in calculus, states that if a continuous function has values that go from positive to negative or vice versa, one value of the function must be 0. Consider moving a straight line across a figure, each successive position of the line parallel to the original one. Make your continuous function the difference in the area on one side of the line from the area on the other side. This difference will sometimes be positive and sometimes negative, so by the fundamental idea there will be a position of the line for which the difference is 0. This argument shows that any region can be divided by one straight line, oriented however you like, into two equal areas.

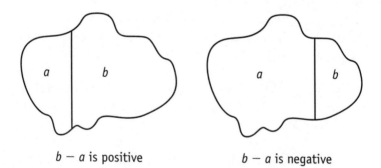

$b - a$ is positive $b - a$ is negative

Applying the result for two lines, each perpendicular to the other, will divide the region into four areas, which will be equal to each other in pairs on opposite sides of the line. If the regions are numbered like the four quadrants of a plane then the equal pairs are I + II equal to III + IV and II + III equal to I + IV. But this is not sufficient for all four pairs to

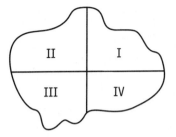

be equal, although a bit of algebra shows that the opposite quadrants must be equal. Subtract, say, II + III = I + IV from each side of I + II = III + IV and you get

$$\begin{array}{r} \text{I} + \text{II} = \text{III} + \text{IV} \\ \underline{\text{II} + \text{III} = \text{I} + \text{IV}} \\ \text{I} - \text{III} = \text{III} - \text{I} \end{array}$$

which rearranges to $2 \times \text{I} = 2 \times \text{III}$, so I must also equal III. Similarly, II will equal IV. Then if I = II or if III = IV, the four quadrants will all contain equal areas.

It is possible to rotate continuously the two perpendicular lines as a unit so that the two separate divisions into equal halves are maintained. The point of intersection will not always be the same during this process, but it is clear that it can return eventually to the original point, at which time the quadrants will all have shifted over one. The same general idea that was used to divide the figure in two can also show that at some time during the rotation two of the adjacent quarters, say, I and II, must be equal in area. Again, the trick is to form a function that is the difference of the two areas in a particular order—say, the function always equals I − II. If in the beginning this number was positive, then the rotation (counterclockwise, as is more common in mathematics) brings I to where II was previously and II to where III was previously. Since we know that I and III were originally equal, the difference I − II must now be negative. Thus, somewhere along the way the point must have passed

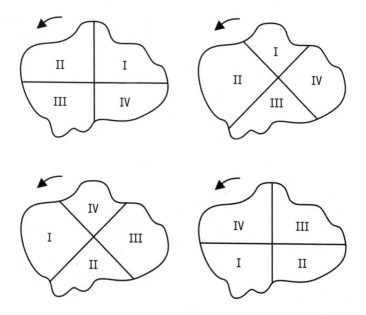

through a region where the difference I − II was 0, so I = II at that point, completing the proof.

This kind of proof is known as an *existence proof*. It shows that the point of division into fourths must exist, but it does not show how to construct it. A few mathematicians and philosophers object to existence proofs and would limit mathematics to constructible points and numbers, but most mathematicians think that this would take all the fun out of the subject. It is true that few existence proofs have any practical value.

One item that, in the United States at least, is usually available primarily in four quarters is butter or margarine. The legend on the wrapper of one of these quarter-pound sticks generally states that 1 stick is equal to 1/2 cup. The stick is frequently marked as well with 8 divisions that are identified as being each equal to 1 tablespoon of butter. Most of us who cook use these measurements without thinking about whether they are true or not, and recipes succeed or fail on other grounds: the actual amount of butter involved in cooking, whether for sautéing mushrooms or as one of the main ingredients in pie dough, is seldom essential. The measurement of butter by the pound converted into capacity, which is another name for volume, seems complex. I was curious and measured 8 tablespoons of water, which indeed is quite close to 1/2 cup. Thus, 4 tablespoons is equivalent to 1/4 cup, a handy comparison to know.

In music one of the important harmonic intervals is the fourth, the interval from C to the next higher F or from G to the next higher C. The number here is not the fraction 1/4 but the ordinal number "fourth," since the interval combines the first note of a scale and the fourth note of the scale (see discussion at **1/5**). The actual interval as a fraction would be either 4/7 of the ordinary scale or 5/12 of the chromatic scale. But the original discovery of Pythagoras of the relation between number and music was that when the ratio of the lengths of two strings of the same general nature is 3 to 4—that is, when one string is 3/4 the length of the other—the shorter string will sound a fourth above the longer and the two strings plucked together will be in harmony. For a single string, divide it at the 3/7 point or the 4/7 point to achieve an interval of a fourth.

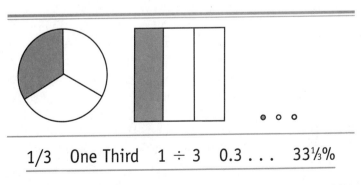

1/3   One Third   1 ÷ 3   0.3 . . .   33⅓%

FIELD MARKS Although 1/3 is in terms of the size of its numerator and denominator the next fraction after 1/2, it is seldom encountered outside cooking. Most other measurements are given in fractions with denominators that are powers of 2, whether for carpentry or for stock quotations.

SIMILAR SPECIES The other third less than 1 is 2/3, which is almost as familiar (or unfamiliar) as 1/3. Because of the way the digit repeats infinitely in the decimal representation, not only does 1/3 convert to a percent that has a fraction of 1/3 in it, but also 2/3 will convert to a percent that has a fraction of 2/3 in it. The same kind of thing occurs with the ninths, for the same reason. The fraction in the conversion of 1/9 to a percent is 1/9, for 2/9 it is 2/9, and so on.

PERSONALITY Notice that if you separate an amount into thirds, one part is exactly half the remaining amount; for example, a third of 51 is 17, which is half of 51 − 17 = 34. Consider a typical probability model such as drawing different-colored marbles from a bag without looking into it. While the probability of drawing a specific one of three different marbles from a bag is 1/3, the odds of getting that marble are 2 to 1 against you, since there are two incorrect marbles and only one correct one. This same idea can often be used to check an answer for 1/3 of an amount, although it is not much help in finding the answer in the first place. But if your calculator tells you that 597 ÷ 3 is 232.33333, you can mentally check by noting that 600 − 230 = 370 and half that is 185, not very near to 232. The correct answer is 199 (note, too, that 597 is almost 600, so 1/2 of that should be near 600 ÷ 3 = 200).

ASSOCIATIONS Most cooks know that a teaspoon is 1/3 of a tablespoon, one of the few places where a person normally encounters 1/3 in measurement. Measuring cups, however, are usually marked in thirds as well as in quarters, partly because that can be useful in correcting the amount of ingredients in a recipe that was originally for, say, six persons that you want to make for two. Then, because the measurements are there, some recipe writers use them, although for most purposes in a recipe 1/4 or 1/2 of a cup could be used where 1/3 is specified. Cooking is not an exact science.

The concept known as charge is one of the central ideas of modern physics. Soon after the electron was first measured in 1897 it became clear that all electrons have exactly the same unit of electromagnetic force associated with each individual particle. That unit came to be assigned a value of −1 (*see* discussion at **Genus Integral, −1**). A dozen years later it became apparent that another subatomic particle, which we now know as the proton, has *exactly* the same but opposite charge, which was assigned the value +1. A score of years after that the first uncharged particle, the neutron, with charge 0, was discovered as well as an antielectron, the positron, with charge +1. As other particles were encountered, each had a charge of one of these three integral values. Although no one really knew why these were the only possible values of charge, it became dogma of sorts that charge comes only in integral units. Thus, when quarks were first proposed in 1964 by Murray Gell-Mann [American, b. 1929] (whose work won him the Nobel prize in physics in 1969) and independently by George Zweig [American, b. 1937] to account for the relationships among various kinds of baryons, other physicists were surprised because the quarks all have charges of 1/3 or 2/3. Quarks cannot be directly observed, however, and always come in pairs or triplets whose charge sums to the long-familiar integral values (*see* discussion at **2/3**). The quarks with a charge of 1/3 actually have a charge of −1/3, while some antiquarks have a charge of +1/3. The main quark in ordinary matter is the down quark, which has a charge of −1/3, but its antiparticle, usually called the down antiquark rather than the antidown quark, is also a constituent of the positive pion (a union of an up quark and a down antiquark), which is an evanescent part of ordinary matter. The ordinary down quark is united with two ups to form a proton and with another down and an up to form a neutron. From this statement, if you know that the proton has a charge of +1 and the neutron a charge of 0, it is possible to deduce that the up quark has a charge of +2/3.

Although the quarks of ordinary matter are the up, the down, and their antiquarks, higher-energy quarks also contain fractional charges. In particular, the quarks known as strange and bottom (once called "beauty," but that name seems to have been abandoned) also have a charge of −1/3, while their antiquarks have a charge of 1/3.

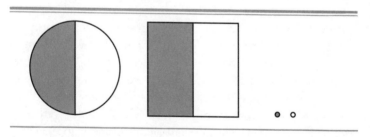

1/2   One Half   1 ÷ 2   0.5   50%

FIELD MARKS   Certainly 1/2 is the simplest and probably the original fraction. Notice that any fraction has two different models that are conceptually different, although we don't always notice the difference. One model of 1/2 is continuous division into two equal parts—slicing the sandwich in half, so to speak. If a figure has a line of symmetry, then one half can be folded over the other, which means that the parts are not only equal in size but also the same shape—congruent, as the geometers say. The other model of 1/2 begins with two equal, but separate, objects, in which case 1/2 is taken to mean one of the two. Kindergarten and first-grade teachers are aware of the conceptual difference between half an object and one out two objects, and they carefully teach that both are indicated by 1/2. A third concept is a meld of the two. If there are an even number of objects, then the objects can be separated into two equal sets, and each set is half the total number. I say this concept is a meld because when the number of objects becomes very great, as in the number of grains of wheat in a bushel, the area concept takes over from the set concept. A philosopher might ask what the dividing line is. There is no answer to this. The question is equivalent to the ancient Greek paradox of the heap—clearly one grain is not a heap of grain, and adding one grain will not make a heap, so at no point can you go from a few grains to a heap by adding one grain at a time.

SIMILAR SPECIES   Often 1/2 is used as a unit, making it more like 1 than like other fractions. In that case also the word "half" in the name is more like a measurement unit, so that we measure 1 half, 2 halves, 3 halves, and so forth. A British drinker who has had 5 "halfs" in a pub has had in some ways a different experience from one who has consumed 2 pints and a half.

PERSONALITY   For most people the operations of halving and doubling are among the easiest to learn, a skill which is often used to advantage by elementary teachers working with the beginning ideas of multiplication. I suspect strongly that both operations, which are in some ways more fundamental than $1/2 \times n$ or $2 \times n$, are at least partly hardwired into the human brain, somewhere near our natural number line, although I do not know any research on this aspect of mathematics. Not only is it especially easy to find half of any number, but even without thinking about it we continually estimate halves and are generally close in such estimates.

Division by 1/2 is the same as multiplication by 2, which should be as easy an idea as halving but is not.

ASSOCIATIONS

The idea of separating any shape whatso-ever into two equal-area regions was discussed to some degree as a part of a proof at **1/4**, but there are many other interesting points to be made about halving objects. One of them, briefly announced at **1/8**, is the ham-sandwich theo-rem, which states that any three shapes placed on three par-allel planes can be separated by a single straight cut into equal areas—the location of the cut may be hard to find, but it is there. In this form the proof is more advanced than we want to contemplate, but a similar version for two shapes that lie in the same plane is not so complex. As noted in the discussion at **1/4**, any shape whatsoever can be separated into equal areas by a line oriented however one likes. Simply move the line along and note the difference between the area on one side and that on the other. Since that difference will be positive at some positions and negative at others, a continuous change implies that it must be 0 at some point; hence at that position the two areas are equal.

Now consider two different regions, each shaped how-ever you wish, in two different parts of the plane. Begin with a line that divides one of them in two but that does not intersect the other. Since this line can be oriented however one wishes, rotate it so that it begins to intersect the second figure, keeping the first figure divided equally. If the rotation is continuous, the difference concept can be applied to the second figure, so both can be divided into equal areas by the same line.

256

The same general idea and reasoning can be used to show that no matter how odd its shape, a plane figure can be divided with a single straight line in such a way that both the area and the perimeter of the figure are halved. The only requirement is that the boundary separates the plane clearly into one part inside it and one part outside—a ring or other figure with holes in it would not work.

All particles have a mysterious characteristic called spin, so called because it behaves in some important ways like angular momentum, the momentum developed by spinning something around a central axis. For one similarity, the type of measure used for spin and angular momentum is the same. A big difference, however, is that angular momentum is a continuous quantity, but spin is measured in units that are given as 0, 1/2, 1, 3/2, 2, and so forth—all multiples, it would seem, of 1/2. While 1/2 seems an odd choice for a unit, it is not even the real value of the basic unit. Every particle spin is really a multiple of the number called Planck's constant, or h, divided by 2π (h is a constant measured at about $1.05457266 \times 10^{-34}$ joule-second with an accuracy of about 0.60 parts per million). Somehow in the history of quantum mechanics, this number became equated to 1/2 (the occasion is that the spin of an electron outside of the orbit of an atom is 1/2 the spin of an electron in orbit; also, people simply dropped the constant factor). The spin is exactly 1/2 for most of the familiar subatomic particles, including the electron, proton, neutron, neutrino, and quark. These are the particles, collectively called fermions, that make up ordinary matter. (The particles that create forces are the bosons.) Fermions get their name because the mathematics of their interactions with each other were worked out by Enrico Fermi [Italian-American, 1901–1954]. Fermions all obey the Pauli Exclusion Principle (named for the German physicist

Wolfgang Pauli, who proposed it in 1925) that no two identical fermions in an atom can have the same set of quantum numbers. The fermions of ordinary matter include electrons, neutrinos, and quarks as well as the proton and neutron that are made from quarks, and there are also higher-energy versions of these particles, including such particles as muons and tauons, which are massive versions of the electron, as well as particles known as strange and charmed (because they contain quarks called strange and charm).

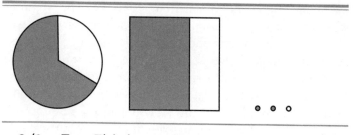

2/3   Two Thirds   2 ÷ 3   0.6 . . .   66⅔%

FIELD MARKS   The Egyptians, who had no truck with most fractions that did not have 1 as a numerator, had a special symbol for 2/3. Other early cultures also treated 2/3 as different from the other fractions, most of which were unit fractions. The Egyptian word for 2/3 has been translated several ways, but the one that seems to come closest to what they had in mind is "both parts." Think of fractions as beginning with 1/2, so that an amount greater than, say, 3 but less than 4 might be called "three and part of four" to mean 3½, while an amount somewhat greater might be called "three and both parts of four" to mean 3⅔. At least this seems on linguistic grounds to be the reasoning, since Akkadian-Babylonian says "two hands" for 2/3, and classical Latin and even modern Greek still say "two parts" (*bes* in

258   Kingdom Number

Latin is a contraction of *binae partes assis*, "two parts of the whole"—the Romans considered unity as divided in 12—and the Greek is quite explicit).

SIMILAR SPECIES

The number that in many ways is paired with 2/3 in most people's mind is 3/4. The two numbers are the fractions in lowest terms with a nonunit numerator that have the smallest numerator and denominator. Probably that is why they both occur so often in problems in textbooks, although perhaps not so often in ordinary life since few things are measured partly in thirds and partly in fourths. The exception occurs in recipes, which often use thirds of a cup and fourths of a cup in the same recipe. It is likely the occurrence in textbooks that has given most people the instantaneous recognition that the lowest common denominator of thirds and fourths is twelfths. People are a bit slower in remembering that 3/4 is slightly greater than 2/3. If they stop to think and convert each to twelfths (2/3 = 8/12 while 3/4 = 9/12) or to percents (2/3 = 66⅔%; 3/4 = 75%) the comparison is easy.

While it should be obvious that 2/3 of 3/4 (two of the three fourths) is 1/2, some people are surprised until they make the computation 2/3 × 3/4. Of course it is less obvious, but equally true, that 3/4 of 2/3 is 1/2.

ASSOCIATIONS

The curve $x^{2/3} + y^{2/3} = a^{2/3}$ where a is some number and x and y are the variables associated with a rectangular coordinate system is a curve that is generally called the astroid for its starlike shape.

The same curve is also known as a hypocycloid of four cusps, which means that it is the curved path followed by a point on a small circle that is rolled around inside a large circle (that's the hypocycloid part—a cycloid is the path of a point on a circle rolled along a line and a famous curve for

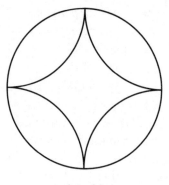

Astroid

other reasons as well) when the small circle is 1/4 the size of the large circle, so that there are four places where the path suddenly changes direction. The same curve results if the small circle is 3/4 the size of the large circle.

The up quark, one of the fundamental constituents of matter, has a charge of 2/3. Like the other quarks, it is never observed by itself but is always combined with one or more other quarks to form a particle with an integral charge. The proton, for example, is two ups for a charge of 4/3 combined with a down that has a charge of −1/3, giving the proton its integral charge of +1. Similarly, the neutron is one up (+2/3) and two downs (−2/3) for a neutral charge of 0.

There are also other quarks that have a charge of 2/3. These are known as charm and top (once sometimes called "truth"). Charm and top occur only in higher-energy particles that are not part of atoms. All quarks also have antiquarks with the opposite charge, so the up, charm, and top antiquarks all have a charge of −2/3.

In the discussion of the CPT theorem (*see* **Genus Integral,** −1) I noted that experiments in 1998 demonstrated that time conservation is violated at the particle level. The connection between time and the actual experiments (there were two of them reported) is difficult to grasp, but one of the experiments is interesting even when time is left out of the picture. Researchers at the CERN particle accelerator complex in Switzerland produced a stream of several million kaons and antikaons and watched them decay. Although the original process produced equal numbers of the two particles, when the decays were measured the probability was almost exactly 2/3 greater that the original antikaon would become a kaon as it traveled through space than that the original kaon would become an antikaon.

Why is the observable universe filled mostly with matter? Theory would suggest instead that the big bang produced a universe filled equally with matter and antimatter. This question is still unresolved, but it is known that the kaon decays into an electron (and a pion) while the antikaon decays into a positron (and a pion). Thus, in a universe filled with kaons, which are the higher-energy equivalent of our ordinary matter's pions, after a time, for every three antikaons there would be five kaons—5 is 2/3 greater than 3. Shortly afterward, when the kaons and antikaons decayed, there would be five electrons for every three positrons. The electrons would, upon meeting the positrons, vanish into a high-energy photon (such as a gamma ray). Since there are more electrons formed in kaon decay than positrons, the result, for those electrons at least, would be that the positrons (which are antimatter) would tend to vanish and the electrons would become a matter residue.

For this explanation to account for the observed matter–antimatter disparity, it needs to apply also to some other high-energy decay processes. New particle accelerators that can examine some of these processes were being built in the late 1990s and in a few years should determine whether or not the asymmetry of time can account for the asymmetry of matter.

Genus *Real*
(Number-Line Numbers)

The Greeks learned that some numbers cannot be measured by any chosen unit—for example, if you choose a unit and make it the length of one side of a square, the diagonal of a square cannot be measured exactly with the same unit (*see* **Irrational Family,** pp. 274–278). The Greeks solved the logical problem that this discovery created by separating the idea of number, which would have included at that time the positive rational numbers only, from the idea of magnitudes, which are the measures of line length (or of area or volume). They called magnitudes that could not be measured by their chosen unit "incommensurables." Some scholars in Europe's Middle Ages were the first to represent such magnitudes as numbers. Today mathematicians recognize that most of mathematics depends on identifying all magnitudes as numbers. These quantities have come to be called the *real numbers*, and they form our genus *Real*.

Richard Dedekind [German, 1831–1916] gave the first logically sound definition of the real numbers, which he conceived on Wednesday, November 24, 1858, far earlier than his 1872 presentation of the formulation that has remained the basis of our understanding of the real numbers today. Before Dedekind's definition mathematicians had implicitly used the idea that a real number is the same as the limit of a sequence of rational numbers. When mathematicians of the eighteenth and early nineteenth century made this idea explicit, however, they employed vague ideas of what a limit might mean or appealed to geometric intuition. Dedekind was not content with such laxity. He recognized that if the limit of a sequence of rationals is not itself rational, for example, $\sqrt{2}$, then the logic of defining a number as such a limit of the sequence seems badly flawed.

Instead of a limit concept, therefore, Dedekind turned to an idea that Euclid had used to explain incommensurables (what we now call irrationals). Euclid, probably borrowing from a work by Theaetetus [Athenian, c. 417–369 B.C.], showed that an incommensurable magnitude, such as $\sqrt{2}$, divides the rationals into two classes, those above the incommensurable and those below. In the 1830s William Rowan Hamilton also proposed this as a definition of irrationals but did not pursue the idea (he was an alcoholic and as often is the case with such people would conceive great ideas but fail to follow them through). Dedekind assumed that the rational numbers form a well-defined set, which means that we can always tell whether or not an entity is a rational number. Furthermore, the set of rational numbers is ordered in such a way that, given any two different rational numbers, you can tell which of the two is greater. Then Dedekind defined a real number as an entity that separates the whole set of rational numbers into two sets, one of all numbers greater than the entity and one of all numbers less than it. He called this entity a cut.

Dedekind's essential argument is not hard to follow, although a detailed proof is necessarily lengthy. Consider two sets of rational numbers, one whose squares are greater than 2 and another whose squares are less than 2. These sets are well defined and clearly exist. But from our earlier analysis of the diagonal of a unit square it is clear that there is a point on the real line, which we call $\sqrt{2}$, that is not in either of these sets. This point is the "cut" that separates the two well-defined sets. Dedekind then specified that the point derived from that cut into two sets is defined as the value of the real number, although he was a little vague about what the point was exactly—clearly he meant it to be a number and not a geometric point.

More recent mathematicians have sometimes completed the definition in another way. They define a real number as the two infinite sets on either side of the cut, being somewhat more comfortable with infinite sets than with operations. Although this resolves the lingering ambiguity of Dedekind's original version, it boggles my mind to think of a single real number such as $\sqrt{2}$ as a pair of infinite sets of rational numbers. This is, however, the closest that anyone has been able to get to the idea that works so well when the integers are defined as pairs of natural numbers; the rationals are defined as pairs of integers; and, as is discussed later, the complex numbers are defined as pairs of real numbers. For this reason, the great imaginative leap in mathematics (as one jumps across the river of number from stone to stone) is between the rational numbers and the real numbers.

But if we forgo the mathematically impeccable and rather impenetrable concept of numbers as pairs of infinite sets and return to the slightly shadier version Dedekind produced, we have an intuitive picture of real numbers that is effective and sound enough for anyone but the most finicky logician. The infinite line of geometry can be taken as the model of the real numbers: for every real number, there is a point on the line; for every point, a real number. In more philosophical moods, I worry that continuity of a line may not correspond to any actual entity—atoms, even when pictured as fuzzy waves, are still separate individuals, and between two atoms there is space. But when it comes to practical mathematics, all but the most ascetic individuals treat the line as continuous and infinite and therefore coextensive with the real numbers.

The other idea that is commonly used with the real

numbers is the association between these numbers and the infinite decimals. Every infinite decimal represents a point on the real number line. A sketch of the proof of this assertion is easy to follow. Suppose that the real number shown by a point on the line is actually an infinite decimal that has the digits 0.*abcde* . . . , continuing as far as one needs to go. Then the first digit 0.*a* means that the number describes a point that is greater than or equal to the rational point *a*/10 and less than 1, localizing the point in that set of rational numbers. Similarly, the first and second digits combined localize the point between the number *ab*/100 and 1, while the second and third localize it between *abc*/1000 and 1. Continuing in this way produces the infinite set of rational numbers that Dedekind tells us localizes a real number, a set that is all less than the cut at 0.*abcede* . . . . Mathematicians say that this is an infinite sequence of rational numbers that approaches the real number "from the left." A second set that approaches "from the right" can be established by looking at the other intervals and representing them as rational numbers. This is easier to show with a numerical example than to explain in the abstract.

Let's take as our number $1/\phi$, which is a real number that begins 0.618033988. . . . (you may recognize the reciprocal of the Golden Ratio here—*see* discussion at **1.61803 . . .**). To approach from the left, the first rational number is 0.6, the second 0.61, the third 0.618, and so forth. To approach from the right, the first rational number is 0.7, the second is 0.62, the third is 0.619, the fourth is 0.6181, then 0.61804, 0.618034, 0.6180340, 0.6180399, and so forth. Notice that the rule is to add 1 to the last digit unless the digit is 9, in which case you change that digit to 0 and add 1 to the digit before it. This approach from the right will produce the second set of rational numbers, all greater than the real number $1/\phi$. With the two infinite sets of rational numbers above and below the cut, the criterion for a real number is met. The same procedure will work for any number that can be expressed as an infinite decimal, so all infinite decimals represent real numbers.

The set of infinite decimals is not exactly the same as the set of real numbers, since the same point on the line is represented by two different decimals, as is especially noticeable for the real numbers that are also natural numbers. The natural number 2, for example, when treated as a real number can be either 2.000 . . . or 1.999 . . . , since both reach the same point on the line. Despite this complication, the infinite decimals make an excellent practical model for the real numbers.

Mathematicians also use other models of the real numbers, depending on the situation. Radian measure of angles is equated with real numbers for many kinds of problems, not all involving angles. Also, real numbers appear in situations involving vectors. But for understanding, nothing beats thinking of real numbers as the points on a line.

Commonly Seen Species

0.110001000000000000000000010000 . . .
The Basic Liouville Number

FIELD MARKS 0.110001000000000000 0000010000 . . . might be called the number that was described only because it is interesting to mathematicians. In 1844 the mathematician and publisher Joseph Liouville [French, 1809–1882] determined that there exist numbers that are not the solution to any algebraic equation. The basic Liouville number is the smallest of the numbers he found that fit that description. It is the sum of all the numbers of the form $10^{-n!}$ as n

goes through the natural numbers from 1 to infinity, that is, the sum

$$\frac{1}{10^1} + \frac{1}{10^2} + \frac{1}{10^6} + \frac{1}{10^{24}} + \frac{1}{10^{120}} + \frac{1}{10^{720}} + \cdots$$

As far as is known, the only use for such numbers is to alert mathematicians to the existence of such numbers.

SIMILAR SPECIES The whole class of Liouville numbers consists of the products of *almost* any other number with the basic Liouville number. Which numbers can be used? Call the basic number L and use a to represent some specific number that is not 0, but otherwise could be *any rational number or even any root of a rational number,* such as the square root or cube root (these roots may themselves not be rational numbers, as is discussed at **Irrational Family,** 274–278). Then aL is a Liouville number.

Notice that there are as many Liouville numbers as there are values for a. The possible values for a are all solutions to some algebraic equation. For example, if a is a rational number, such as 4/7, then it is easy to write the equation $7x = 4$ that has a as its solution. On the other hand, if a is a square root, such as $\sqrt{5}$, or some other kind of root, such as a fourth root, then it is also easy to write equations such as $x^2 = 5$ or $x^4 = 5$ that have a as the solution. Since each value for a can be matched with a Liouville number, there are at least as many numbers that are not solutions to algebraic equations (numbers of the form aL) as there are numbers that are such solutions (the various values for a). Notice that it is necessary to say "at least as many" because in addition to the Liouville numbers, there may be other numbers that are not solutions to any equations. And, as you may already know, this is in fact the case.

Another easily written number that is not the solution to any algebraic equation is formed by writing the Hindu-Arabic representations of the natural numbers in order as a decimal fraction:

0.1234567891011121314151617181920212223232425 . . .

PERSONALITY Other Liouville numbers may or may not be easily spotted should one suddenly run into one. If $a = 7$, then $aL = 0.7700070000000000000000070000 \ldots$, which displays the hallmark of a Liouville number right away—the appearance of long strings of 0s that rapidly become longer as one moves farther to the right of the decimal point. But recognition becomes more difficult if a is something other than a small integer. For example, to the first fifteen digits of aL when a is equal to 4/7, the Liouville number is 0.0628577142857143. An astute number-watcher might spot the hidden seventh from the part that has 142857, which is the six-digit signature of a fraction with a denominator of 7. But one would need to go many more digits before the essential nature of the number would appear.

Algebraic Family

The achievement of Liouville was to produce a set of numbers that are not the solutions to any equation that does not already contain such a number—clearly if L is a Liouville number, the equation $x = L$ has a Liouville number as its solution. Numbers that are the solutions to equations are known as *algebraic numbers*. But to make this statement meaningful as a definition of a particular set of numbers, we have to specify rather exactly what kinds of equations will be included.

An equation in this discussion is any mathematically grammatical statement of equality involving a variable and some numbers. Solving an equation means finding the set of numbers that can replace the variable in the equation to produce a sentence that is always true. We may begin by taking all the numbers involved in the statement of the equation to be natural numbers; this or some similar restriction is needed to make any sense of equation theory. Oddly enough, the restriction to natural numbers can for the most part be lifted after all the results are in—that is, results obtained with the restriction apply to almost all kinds of numbers.

Next, we need a way to classify equations. We will consider only equations formed by the operations of addition and multiplication, which includes natural-number powers of the variable, since such powers are a shorthand form of multiplication. The simplest way to group equations is by the highest power of the variable used in the equation, a natural number called the degree of the equation. Since $x^1 = x$, equations such as $2x + 4 = 0$ or $3x + 1 = 5$ are counted as first-degree equations. If you solve either of these equations, you leave the natural numbers immediately, however, for the number for x that will make $2x + 4 = 0$ a true statement is -2. Similarly, the number for x that makes $3x + 1 = 5$ a true sentence is 4/3, a rational number. So even if the requirement is to stick to natural numbers for everything else, the values that work for x may be some other kind of number.

The situation becomes more complicated if you raise the degree to the next natural number, 2. First, it appears that more than a single number can be a solution: an equation such as $x^2 = 9$ has two correct solutions, since both $3^2 = 9$ and $(-3)^2 = 9$ are true. The situation gets more complicated if you consider $x^2 = 2$ or $x^2 + 1 = 0$, for the solutions to both of these simple-appearing equations are neither natural numbers, integers, nor rational numbers. The solution to $x^2 = 2$ is a real number, and the solution to $x^2 + 1 = 0$ is yet another kind of number, to be explored at **Genus Complex.**

So it is clear that the process of solving equations can greatly enrich the kinds of numbers that are known to exist. Even when the numbers used in the equations are kept to natural numbers and even for the simple operations of addition and multiplication only, equations of the first degree lead to numbers of all the types considered up to now, and equations of the second degree produce two entirely different kinds of number. So it might be thought that proceeding in this same way to equations of higher and higher degrees might reveal all sorts of numbers.

It turns out that nothing of the sort happens. As mathematicians learned to deal with equations of degrees 3 and 4, they found that there were ways to show that the solutions could be found using equations of lower degree. So no new kinds of number emerged. A series of mathematicians between 1799 and 1830 showed, however, that the equations of degree 5 and higher cannot always be solved by using lower-degree equations. Thus, one might possibly anticipate finding yet some other kind of number as the solution to an equation based on natural numbers if the degree were sufficiently high.

Another result derived during that time—the fundamental theorem of algebra—shows that no new numbers can emerge at higher degrees. Gauss proved this theorem four different ways at different times in his life, first in 1798. The essence of the fundamental theorem is that every equation formed by an equation based on a finite amount of addition, multiplication, and powers has at least one solution. One consequence of the fundamental theorem is that, even when you let the numbers involved in writing such an equation be any of the various kinds of numbers that result from solving equations through the second degree, that solution will be one of those same kinds of numbers. This is as far as one can go with algebra.

It would make sense to call all these numbers "algebraic" and any numbers that might come from some other source "nonalgebraic." Although mathematics enjoys a reputation for clarity and precision, it does not usually live up to it where terminology is concerned. When mathematicians use the term "algebraic" to refer to a number, they usually do not mean to include the ones that are like the solution to $x^2 + 1 = 0$. Such numbers have their own name, "imaginary," as opposed to numbers that have no imaginary parts, which are called "real." Careful mathematicians recognize that what are usually called "the algebraic numbers" include only the real algebraic numbers and specify that. But not all are that careful, so remember, the algebraic numbers are usually the real ones only and not any of the imaginary types.

1.414213562373 . . . $\sqrt{2}$
The Square Root of 2

FIELD MARKS   The square root of 2, or $\sqrt{2}$, is the square root of the smallest natural number that is not a perfect square, so it is often taken as the archetype of such numbers. Note that $\sqrt{2}$ is an example of what modern mathematicians call a *radical*, indicated by the *radical sign* $\sqrt{}$; the number 2 as shown in the symbol $\sqrt{2}$ is the *radicand*. It is easy to see that radicals for which the radicand is a natural number, such as $\sqrt{2}$, are all algebraic numbers, since they are all solutions to equations of the form $x^2 = n$. It is less obvious that when the radicand is not a perfect square such radicals are not rational numbers (*see* **Irrational Family,** pp. 274–278). It is important to note that as written $\sqrt{2}$ is a positive number. Even though 2 has two square roots, the symbol $\sqrt{\ }$ by itself is reserved for the positive one. To get the negative root, you need to write $-\sqrt{\ }$.

SIMILAR SPECIES The number $\sqrt{2}$ is often taken as the exemplar for algebraic numbers that are not rational, since it is easy to understand and can easily be proved not to be a rational number, but it is an algebraic number as one solution to $x^2 = 2$ (the other is $-\sqrt{2}$). All other radicals that do not have perfect-square radicands fit the same mold—they are algebraic but not rational. Furthermore, cube roots that are not the roots of perfect cubes, fourth roots that are not the roots of perfect fourth powers, and so forth are all in the same boat—algebraic but not rational. The kind of root that a radical represents is shown by a small number called the *index* that is written inside the "check" of the radical sign, so $\sqrt[3]{8}$ is the cube root of 8, which is 2 (although along with this familiar real cube root of 8, there are two additional cube roots that are not real).

Mathematicians sometimes work with a system of numbers of the form $a + b\sqrt{2}$, where a and b are rational numbers. Such a set of numbers has the advantage of behaving like the full set of rational numbers even though, of course, none of the numbers involved is rational unless $b = 0$. The occasion for using this kind of system is not so much to accomplish anything practical, but more to illustrate how the structure of a number system works.

PERSONALITY Like $\sqrt{2}$, the square root of 3, or $\sqrt{3}$, is a commonly occurring radical that often results from the simplification of other numbers. The square root of every even number that is not also a multiple of 4 and of every multiple of 3 that is

Kingdom Number

not also a multiple of 9 will have buried within it a factor of $\sqrt{2}$ (for the even numbers) and $\sqrt{3}$ (for the multiples of 3). The reason is that for any two numbers a and b the product $\sqrt{a} \times \sqrt{b} = \sqrt{ab}$, so a number such as $\sqrt{150}$ can be thought of as

$$\sqrt{150} = \sqrt{2 \times 3 \times 25} = \sqrt{2} \times \sqrt{3} \times \sqrt{25}$$
$$= 5\sqrt{2}\sqrt{3}$$

Because $\sqrt{2}$ and $\sqrt{3}$ occur so often in mathematics, I always encourage students to memorize at least the first few digits of their decimal expansion, which are typically given as 1.414 and 1.732. Then these roots can be used to evaluate many other radicals—for example, $\sqrt{2} \times \sqrt{3}$ is $\sqrt{6}$, so a good estimate for $\sqrt{6}$ is 1.4 × 1.7, more easily pictured as 0.7 × 3.4, which is about 0.7 × 3.5. Now think of 35 as 7 × 5, so this becomes a decimal version of 7 × 7 × 5 = 49 × 5, which is 5 less than 250—so $\sqrt{6}$ is about 2.45. A three-place table gives the value as 2.449, so 2.45 is not bad for a rough approximation.

ASSOCIATIONS The numbers $\sqrt{2}$ and $\sqrt{3}$ are familiar in part because they emerge in any consideration of the two fundamental right triangles. One of these is the isosceles right triangle and the other is the right triangle that emerges as half of an equilateral triangle, produced by an altitude to one of the bases. If any one side of an isosceles right triangle is taken as a unit, then $\sqrt{2}$ will appear elsewhere; similarly, if one side of the half-equilateral triangle is taken as the unit, then $\sqrt{3}$ is present. This is related closely to common trigonometric values:

$\sin 45° = \cos 45° = \dfrac{\sqrt{2}}{2}$ and

$\sin 60° = \cos 30° = \dfrac{\sqrt{3}}{2}$; $\tan 60° = \sqrt{3}$; $\tan 30° = \dfrac{\sqrt{3}}{3}$

The paper sizes that are standard in England are designed so that each of the eight A sizes (0 to 7) can be halved to produce a sheet similar to the original in ratio of length to width. If one takes the middle of the sequence, A4, the measurements are 210 mm by 297 mm, which translates to 8.27 inches by 11.69 inches, rather close to the current American standard letter size of 8.5 by 11 inches. It seems striking that the ratio of 297 to 210 is 1.4142857, which is $\sqrt{2}$ to the thousandths place and just one digit away from being $\sqrt{2}$ to the ten-thousandth place. The other paper sizes maintain about the same ratio as well. But this is not some deep mathematical property. If you start with some other small size, such as the familiar 3 by 5 (inch) index card, you can set up the same kind of system by taking the small side of the next larger size to be the same as the large side of the original—that is, if 3 by 5 is P_0 then the short side of P_1 will be 5. Now the ratio of 5 to 3 is 1.66666 . . . so the long side of P_1 will need to be 5 × 1.66666 . . . = 8.333333 . . . , which you round to 5 by 8, the obvious size for the next larger index card. The following one, P_2, will have 8 for its smaller side and an approximation of 13.3333 . . . as the longer one, giving an 8 by 13 index card.

Irrational Family

Ironies abound in the history of the discovery of irrational numbers by an early Pythagorean mathematician. Not only was the basis of the Pythagorean philosophy the primacy of natural numbers, which were viewed as the foundation and guiding principle of the universe, but also it appears likely that the discovery that not all geometric lengths could be expressed in terms of natural numbers was made while contemplating the symbol of the Pythagorean society, the pentagram. A careful examination of original Greek sources made in 1987 by D. H. Fowler in his book *The Mathematics of Plato's Academy* shows that early references to this discovery could have as easily referred to the pentagram as to the square, which is often given as the geometric figure involved in overthrowing the rational numbers.

The pentagram (or pentacle) is the five-pointed star (*see* discussion at **Genus Natural, 5**). It provides an early example of a fractal, since repeating the simple operation of drawing diagonals in pentagons produces a series of self-similar figures. Fractals were first specifically discussed by Benoit Mandelbrot [Polish-American, b. 1924] in 1975. They are technically described as curves with an intrinsic dimension higher than the dimension of the space in which they are embedded, but their main characteristic is self-similarity at various magnifications. For the pentagram, the self-similarity begins with the observation that connecting the interior vertices of the five-pointed star will produce a smaller star that is similar to the first.

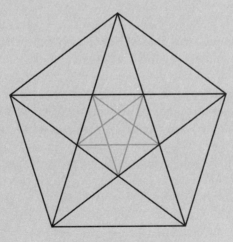

Self-Similarity in a Pentagon

Each smaller star will also include a regular pentagon within it, outlined by the diagonals. This operation can be repeated to produce other stars as many times as you like. Suppose that you choose as a unit of measurement the length of the diagonal that forms one side of a star—or any other length within a particular level of the diagram. Then that unit will be too long to measure the same part of the star *within* the star you chose for the length. This process can also continue as far as you like. Thus, no matter how small a unit you use, it will not be small enough to measure all the stars that you know can be made to exist in the diagram.

When one length can be used to measure exactly another length, the two lengths are said to be commensurable. This argument suggests that parts of the successive stars are incommensurable lengths. But the argument probably would not have satisfied the Greeks, although it may have suggested to them that they analyze the figure further. Aristotle alludes to a proof that some numbers are incommensurable that is based on odd and even numbers. There exists such a proof for the pentagon, and it is easily within the power of Greek geometry. But most historians assume that Aristotle was thinking of a similar proof showing that the side of a square is incommensurable with its diagonal. Today this is usually proved using the Pythagorean theorem along with a side proof that if a perfect square is a multiple of 2, so is its square root. The side proof follows from the idea that the only possible choices for the square root are odd and even (since we began with a perfect square), and of these only the even × even will produce an even perfect square.

Consider a square that has a side of 1 unit and the diagonal of such a square, whose length we will label d:

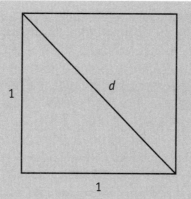

Since the angle between the unit sides is 90°, by the Pythagorean theorem $d^2 = 1^2 + 1^2 = 2$, so $d = \sqrt{2}$. Assume that $\sqrt{2}$ is rational. Then it can be expressed as the ratio of two natural numbers p and q so that the fraction p/q is in lowest terms (p and q have no common factor). Then we can proceed as follows:

$$\sqrt{2} = \frac{p}{q}$$

$$2 = \frac{p^2}{q^2}$$

$$2q^2 = p^2$$

This shows that p^2 must be an even number, since it is equal to $2q^2$, which is clearly even. Therefore, by the side proof discussed earlier, p must have been even to start with. Now replace p with $2k$ since we know p is even:

$$p^2 = (2k)^2 = 4k^2 = 2q^2$$
$$2k^2 = q^2$$

The relationship $2k^2 = q^2$, for the same reasons as before, shows that q was also even to begin with, even though we specified that p/q was in lowest terms. Since this is a contradiction, we know that $\sqrt{2}$ cannot be a rational number, or, in more Greek terms, the side of a square is incommensurable with the diagonal.

$$\frac{A}{B} = \frac{A + B}{A} = \phi$$

1.61803398 . . . ϕ $(1 + \sqrt{5})/2$
The Golden Ratio

FIELD MARKS Many common constants are known by symbols instead of their Hindu-Arabic numeral. The symbols are usually Greek or Roman letters, such as π or e, which are perhaps the best known of such constants. Just behind π and e in general recognition as a special number is ϕ (the Greek letter phi), which represents a concept familiar to the ancient Greeks. They did not call it ϕ, however, any more than Greek geometers called the ratio of the circumference to the diameter π. The symbol ϕ was assigned to this number in the twentieth century on the basis that 1.61803398 . . . is a number that was used in design work by the Greek sculpture Phidias—an idea that has been called into question by mathematics historians. Some mathematics books use the Greek letter τ (tau) for this number instead of ϕ; I do not know the derivation of this practice.

The number ϕ originated in a question that would occur only to the mathematically minded: "Is it possible to divide a segment of a line into two parts such that the length of the whole segment is to the larger part in the same ratio as the length of the larger part is to the smaller one?" The answer is "Yes, provided that the ratio is about 1.61803398, which is known as φ." The situation that leads to this relationship was known as early as Euclid, who introduces it as Definition 3 of Book VI of the *Elements*. Although the relationship is explicit in Book VI, the same idea occurs in a number of places in the *Elements*.

The traditional name of this ratio derives from the old-fashioned language used in discussing ratio and proportion, "division in extreme and mean ratio," which simply means dividing a line so that the relationship of the first extreme to the first mean is the same as that of the second mean to the second extreme. If a proportion is expressed as $A : B ::$ $(A + B) : A$, the language becomes clearer; the outside pair are the "extremes" and the inside pair are the "means" (*see* **Ratio and Proportion,** pp. 244–246).

Today historians of mathematics think that it is likely that the whole question of division in extreme and mean ratio arose in conjunction with the discovery of irrational numbers by Pythagoreans as they contemplated the pentagram (*see* **Irrational Family,** pp. 274–278). When a pentagram is inscribed in a regular pentagon, the diagonals that form the star cut each other in the ratio φ.

By choosing different parts of the line segment to have the value of 1, it is possible to analyze the relationship algebraically. When the two parts of the line are called A and B, the whole line is $A + B$ units long. Let A be the larger part. Then the equation

$$\frac{A}{B} = \frac{A + B}{A}$$

describes a line for which the ratio of the larger to the smaller part (*A/B*) is the same as the ratio of the whole is to the larger part.

For example, let *B* be taken as 1 so that the first ratio *A/B* becomes *A*/1 = *A*. Then the defining equation contains just one variable, *A*, making it $A = (A + 1)/A$. Multiply both sides by *A* to get $A^2 = A + 1$, or

$$A^2 - A - 1 = 0$$

This is a quadratic equation. All such equations can be solved. In this case the solution, which can be found using first-year algebra, is $A = (1 \pm \sqrt{5})/2$, which is the basic value of ϕ (except that one must discard the negative $\sqrt{5}$ because that would make ϕ negative, which as the ratio of two positive numbers it cannot be).

SIMILAR SPECIES   A little more algebra easily shows that the reciprocal of *A* is *B*, and therefore the reciprocal of *B* is *A*. Furthermore, the decimal equivalents are *A* = 1.61803398 . . . and *B* = 0.61803398 . . . , so *B* is exactly 1 less than *A*. Indeed, ϕ is the only number among all the real numbers for which the reciprocals are exactly 1 unit apart.

PERSONALITY   My friend Jim Davis, a painter, pointed out to me that $\phi^2 \times 6/5$ is π to four decimal places, which he says, referring to the Greek construction boom, is "good enough for temple work."

ASSOCIATIONS   Euclid used division in extreme and mean ratio not only in the construction of a regular penta-

gon, but also in constructing the icosahedron and dodecahedron (a polyhedron with pentagonal faces). Later Greek
geometers showed that if the radius of a circle is divided
into the ratio, then the larger part of the radius is the same
size as the side of an inscribed decagon.

It is not known who first recognized that ϕ is the limit of
the ratio of consecutive terms of a Fibonacci sequence (*see*
Fibonacci and Lucas Families, pp. 120–122), but a statement of the relationship first appears in mathematics history as an annotation on a copy of Euclid's *Elements* that
dates from the 1500s (more than 250 years after Fibonacci
published his problem that leads to the sequence). To
arrive at ϕ as the limit you need to look at the ratios of
second to first, third to second, fourth to third (otherwise
you get $1/\phi$). It is easy to calculate the first few ratios and
see how quickly they converge. The first eight ratios as
common fractions are 1/1, 2/1, 3/2, 5/3, 8/5, 13/8, 21/13,
34/21; as decimal fractions these ratios are 1, 2, 1.5, 1.67,
1.6, 1.625, 1.6154, 1.61905, which are easily seen to be
heading toward ϕ, which is about 1.61803. The proof,
however, is a bit harder than these calculations.

The number ϕ has been known as the Golden Ratio
because of analysis of works of art in which ϕ either
appears or almost appears or seems to appear (the Greeks
called it simply "the section"). Kepler, who knew of the
relation of the Fibonacci series to ϕ and who was also a
number mystic, makes a passing reference to ϕ as "the
divine proportion" in relationship to his work connecting
the orbits of planets to the Platonic solids (*see* discussion at
Genus Natural, 5). Early in the nineteenth century German writers on art began to speak of ϕ as the Golden Ratio

or Golden Section (or equivalent terms) and to attribute the beauty of various buildings to its use. Since then many writers have picked up this idea and claimed, without much evidence, that the Greeks used φ to design temples such as the Parthenon. Indeed, some have traced the use of φ in design to the Great Pyramid of Cheops. In the twentieth century some modern architects, notably Le Corbusier, claimed to have used φ in design, and some psychological experiments were interpreted as meaning that people prefer to look at rectangles whose sides are in the Golden Ratio. Painters, including Leonardo da Vinci, Georges Seurat, and Juan Gris, are supposed to have based works on the Golden Ratio, although there is strong evidence that at least in the case of Seurat and Gris this is not the case.

The rectangle whose sides are A and B with a ratio of φ is called a Golden Rectangle. If you place a series of such rectangles inside each other so that they alternate between having the long and short sides as the base and draw a diagonal in each, the points of intersection of the diagonals will lie on a spiral. The type of spiral is known as the logarithmic spiral (*see* discussion of *logarithms* at **2.71828** . . .) or the equiangular spiral. It is equiangular because the angle formed by any line from the origin of the spiral to the curve always makes the same angle with the tangent to the curve at the point where the line meets the spiral. There are several ways to relate this spiral to logarithms; all are somewhat beyond the scope of this book. The relationship of φ to the logarithmic spiral is also involved with the connection between φ and the Fibonacci sequence. The spirals seen in the seeds of a sunflower, for example, which are numerically connected to the Fibonacci numbers, are all logarithmic.

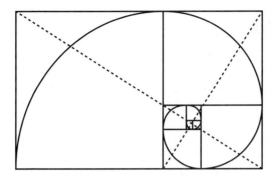

Logarithmic Spiral

2.718281828459045 . . . Euler's Number
e Base of the Natural Logarithms

FIELD MARKS Most people first learn *e* as the number that is the base of the natural logarithms. While this explanation of *e* is true, it is not very helpful, even when one has some general idea of what logarithms are. Here is a primer: a logarithm is most easily understood as an exponent that is used with a particular base to give a specified power. This may be difficult to picture as an abstract idea, so consider an example. If the base is 2, the logarithm 5 will produce the power 32, since $2^5 = 32$. In logarithmic notation, this becomes $\log_2 32 = 5$. Often the base is not specified because everyone using logarithms in a particular field is assumed to know what the base is. In information theory the base is always 2, so information theorists write $\log 32 = 5$. In most of elementary mathematics, however, the base is either 10 or *e*. Engineers and some mathematicians deal with possible confusion by using two different abbreviations for "logarithm." They reserve the abbreviation "log" for logarithms with a base of 10—for example, log 1000 = 3—and use the abbreviation "ln" (for "logarithm natural") for logarithms with the base e, so that $\log_e 10$ is usually written as ln 10. The number ln 10 is the exponent such that e raised to that power is 10, an exponent that is approximately 2.3026. Thus, $e^{2.3026}$ is about 10—the actual logarithm in this case is irrational.

Although the logarithms to base e are most commonly seen abbreviated "ln," you should be warned that in some mathematics works, especially older or higher-level works, the abbreviation "log" is used for the base e, since that is the natural base that emerges from any work with calculus. There was no special need for logarithms to the base 10 except for calculations using tables, which are simpler when the base of the numeration system being used is the same as the base of the logarithm. But modern hand-held calculators and computer spreadsheet programs all use "log" for common logarithms (that is, to the base 10) and use "ln" for natural logarithms (that is, to the base e).

There are many ways to define e. One of the best known is to say that e is the limit of the function $f(n) = (1 + 1/n)^n$ as n increases without bound. A somewhat intuitive explanation of the origin of this function can be given in terms of compound interest, since compound interest can be calculated from the same formula as the function. For an interest rate of 5% compounded annually, at the end of one year one dollar becomes $1 + (0.05 \times 1)$ dollars = $1.05. But if interest is compounded twice a year the effect of the rate during each period is halved, since the rate applies for only half a year. Thus, at the end of the first six months the amount is $1 + (0.025 \times 1)$ dollars. This amount, $1.025, becomes the starting value for the second six months. By the end of that second six months, the total amount is $1 + (0.025 \times 1)$ dollars $+ 0.025 \times [1 + (0.025 \times 1)$ dollars] = $1.050625 (instead of the $1.05 of simple interest). It is also possible and easier to compute the interest after both six-month periods as $(1.025)^2$. If you rewrite the decimal 0.025 as the fraction 1/20, then the formula for one year at 5% compounded twice becomes $(1 + 1/20)^2$ and for ten years at 5% compounded twice each year becomes $(1 + 1/20)^{20}$. This looks like the function whose limit is e. Compounding the dollar at 5% every three months, or four times a year, for ten years would produce $(1 + 1/40)^{40}$. Since e is the limit of this function, the dollar would become $2.72 to the nearest cent after 10 years if compounded every instant. That is, the limit of $(1 + 1/n)^n = e$.

There are many other limits that lead to e, although since e is a transcendental number, like π, there is no simple algebraic representation. One such limit that is an infinite-series version of the function for compound interest is

$$e = 1 + \frac{1}{1!} + \frac{1}{2!} + \frac{1}{3!} + \cdots$$

Another is that e is the value of x for which the xth root of x is a maximum. The value 1.414 to approximate $\sqrt{2}$ has been previously encountered. It does not take much to see that the fourth root of 4 is equal to $\sqrt{2}$, so it is also about 1.414. The cube root of 3 is about 1.442, a little larger than $\sqrt{2}$. For values larger than the fourth root of 4 the numbers are all smaller than 1.414 as the xth root of x tends irregularly toward 1. For example, the 327th root of 327 is about 1.018. Thus, the maximum value of the xth root of x must be somewhere near 1.442, the largest value for an integral x. Indeed, the eth root of e is about 1.445 and any other xth root of x is smaller.

Even if you understand the relationship of e to logarithms and can compute it as a series or by some other method, you may still feel unsatisfied. Is there a simple meaning for e like the simple definition of π as the ratio of the circumference to the diameter of a circle? The answer is "sort of." The geometric meaning of e is that $y = e^x$ is the curve whose slope equals e^x for any point on the curve. (If you know calculus, you will recognize that this means that e^x is the function for which $f(x)$ is equal to its own derivative.)

SIMILAR SPECIES

Another number generally represented by a symbol instead of being written in the Hindu-Arabic system is π. Although both e and π have an odd way of popping up when least expected, some would give e credit as the most interesting number. Others might favor π. Mathematicians from the University of Missouri were responsible for a "top 10" list (à la David Letterman) that comes out strongly for e: "Top $\ln e^{10}$ Reasons Why e Is Better than π." I have made minor editorial changes.

10. e is easier to spell than pi.

9. $\pi = 3.14$ while $e = 2.718281828459045$.

8. The character for e can be found on a keyboard, but π sure can't.

7. Everybody fights for their piece of the pie.

6. $\ln \pi$ is a really nasty number, but $\ln e = 1$.

5. e is used in calculus while π is used in baby geometry.

4. "e" is the most commonly picked vowel on *Wheel of Fortune*.

3. e stands for Euler's Number; π doesn't stand for squat.

2. You don't need to know Greek to be able to use e.

1. You can't confuse e with a food product.

Reason number 3 sounds great, but it is technically incorrect. Most mathematics historians believe that e did not originally stand for Euler, although because Euler made so much use of it, the connection was later made that e is Euler's Number. Furthermore, π is the first letter of the Greek words for *perímetron* and *periphéria,* both of which were used to mean the circumference of a circle, which is the same as the circle's perimeter (I have never quite understood why there are two different words for the distance around a circle; but there are many mysteries in life).

PERSONALITY   Think of e as being near 3 for most purposes or as 2.7 if a more accurate value is needed, which is seldom. There are no rational numbers that are conveniently close to e, as 22/7 is conveniently close to π, although certainly one might use 27/10, which is the same as 2.7. A good rational approximation for e to several decimal places that is better than 2,718,282/100,000 is hard to find, but 2721/1001 is correct to six decimal places and somewhat simpler. The fraction 87/32 is the best one can do with only two digits both for the numerator and the denominator, while the easy-to-remember 878/323 is probably best overall, since it is correct to four decimal places.

On a calculator, e is not a separate key as π is, but better calculators all contain a key for e^x, so a value to some ten digits can be found by calculating e^1.

ASSOCIATIONS   All relationships that grow or decay over time (t) according to a fixed rate follow an equation that is of the form $y = Ae^{kt}$, where A and k are constants determined by the particular situation. This equation describes growth when k is positive and decay when k is negative. It is easier to follow when restated in functional notation, as

$f(t) = Ae^{kt}$. The symbol $f(t)$ that replaces y has the advantage that it shows both that the relationship depends on changes in t and also that for some specific t, such as 0 or 1, you can replace the t with the number to represent the value for that t. Time at the start of the observation ($t = 0$) is $f(0)$; if t is time in ten-year periods, $f(1)$ is the time after the first decade.

Notice that when $t = 0$, the formula becomes $f(0) = Ae^0$ or, since $e^0 = 1$, $A = f(0)$. Thus, the growth or decay function can be rewritten as the functional equation $f(t) = f(0)e^{kt}$. In many calculations the constant k is found to be most easily expressed as a natural logarithm. For example, if you measure the population of a village at 1000 on your first visit and then a decade later find it to be 1500, the growth equation can be simply calculated if you measure time in decades, so that $f(1) = 1500$. The amount at the beginning is $f(0) = 1000$. Combining these with the functional equation (and using 1 for t) produces $1500 = 1000e^k$ or $e^k = 1.5$. This equation is true if k is the natural logarithm of 1.5, often written ln 1.5. Thus, the growth equation can be stated as $f(t) = 1000e^{t \ln 1.5}$, which can then be used to find any population in terms of decades or even years; $f(0.1)$ is the population one year after the first visit.

When k is negative, the equation describes decay instead of growth. Decay equations are often used in radioactive dating methods, since radioactive elements decay according to the general formula used here.

One of the most commonly encountered curves in nature is the catenary, most easily observed when a flexible cable or chain is suspended from two points so that it falls into a curve. This curve looks rather like a parabola, although it is not. If the chain is weighted equally by a series of horizontal weights, as in a suspension bridge, the catenary is reshaped into a section of a parabola. (It may seem that the equal weights of the chain itself would count, but it doesn't work that way. Try hanging two weights of the same size from a suspended chain at equal distances from each end of the chain and you will see the difference.)

The surface formed by revolving a catenary about its axis (a catenoid) is the surface of minimum area, which means that catenaries appear in soap bubbles. Despite its ubiquity, the catenary was not recognized as a separate curve until 1690, when the Bernoulli brothers investigated it. The formula can be stated in several ways, but our interest here is in the version

$$y = \frac{a(e^{\frac{x}{a}} + e^{-\frac{x}{a}})}{2}$$

which clearly shows the relationship to e (*see also* discussion at **Genus *Rational*, 2/3**).

The English physicist Robert Hooke [1635–1703] was the first to note (in 1675) that a catenary makes an ideal arch for carrying a horizontal beam, although he did not know the formula for it, which had not yet been discovered. Hooke simply described it in terms of a chain suspended from its ends and hanging free.

The number e appears in an odd way as part of a probability problem, which can be formulated in many different ways—my favorite is n drunken sailors stumbling into their bunks at random. The probability that one of the sailors has accidentally found the correct bunk is 1 − 1/e, a number close to 0.6321 and surprisingly a bit better than 1/2. A similar surprise occurs if you calculate the following procedure: Choose real numbers at random from the interval from 0 to 1 inclusive and add them. Continue until the sum of the selected numbers is greater than 1. How many numbers will you have to choose in all likelihood before this happens? The answer is *e* numbers as an expectation, although of course sometimes the numbers will be 2, 3, 4, or some other natural number and never *e* exactly.

Transcendental Family

Earlier I defined algebraic numbers as any real numbers that are the solutions to equations formed by adding and multiplying natural numbers (considering the natural-number powers of the variable to be a shorthand form of multiplication) for a finite number of times. That family of numbers clearly includes all the natural numbers, rational numbers, and numbers represented by radicals. Later the real numbers were defined primarily as the numbers that are in one-to-one correspondence with the points on a line or, somewhat more loosely, as any number that can be represented by an infinite decimal—recognizing that many real numbers can be represented by two different infinite decimals. When Liouville demonstrated that there are numbers that are not algebraic, he showed in essence that there are points on the real line that cannot be represented as solutions to algebraic equations. It is these numbers that form the Transcendental Family: they are transcendental in that they rise above the experience of ordinary algebra. In Euler's words, "they transcend the power of algebraic method."

The history of transcendental numbers begins well before Liouville. The Greek problem of squaring the circle—finding a square with the same area as a given circle—by means of straightedge and compass had never been solved. This was one of the three problems left unsettled from Greek geometry (the others are trisecting an angle with straightedge and compass and doubling the cube—finding the side of a cube whose volume is exactly equal to a given cube). Plato had decreed that all constructions be made with a straightedge and a compass that collapsed when lifted off the surface. The three classic problems had been solved with various mechanical means but defied all efforts at construction with Plato's restriction.

In the early eighteenth century, however, mathematicians knew of the relationship between geometry and algebra and could begin to think of squaring the circle as a problem of number instead of solely of geometry. In 1737 Euler showed that e and e^2 are both irrational, and Johann Heinrich Lambert [German, 1728–1777] proved that π is also irrational. The irrationality of π did not, however, resolve the problem of squaring the circle since many irrational numbers could easily be constructed with straightedge and compass—after all, the Greeks knew that $\sqrt{2}$ is both the length of the diagonal of a unit square and what we now would call an irrational number. In 1744, a hundred years before Liouville, Euler began to make a distinction between algebraic and transcendental numbers, although he was unable to show that transcendental numbers actually exist. Adrien-Marie Legendre [French, 1752–1833] also worked on π, demonstrating in 1794 that π^2 as well as π is irrational. He was the first to propose—although not to prove—that the circle could not be squared because π might not be the solution to an algebraic equation—as we would say today, because π is transcendental.

After Liouville showed in 1844 that some transcendental numbers can exist (*see* discussion at **0.11000 . . .**) the next big step came when Charles Hermite [French, 1822–1901] showed that a number that everyone already knew is also transcendental. It is e, which Liouville had also tackled but had been unable to pin down. Finally, in 1882, Ferdinand Lindemann [German, 1852–1939] obtained a place in the history of mathematics by proving that π is transcendental and therefore that the circle cannot be squared by compass and straightedge.

Given all the difficulty in establishing a few transcendental numbers, it is tempting to think that they are very rare. As mathematics is understood today, however, this is far from the case; there have to be many, many more transcendental numbers than algebraic ones. For the explanation, it will be necessary to travel to **Kingdom Infinity.**

3.14159265358979 . . .
π Archimedes' Constant

FIELD MARKS The number symbolized by the Greek letter π is a fundamental constant, often approximated by 3.14 or by 22/7 (note that these are two different approximations as $22/7 \neq 3.14$). The formula $C = \pi d$, which means "circumference of a circle equals π times diameter" provides the fundamental definition of π as the ratio of the circumference of a circle to its diameter. Often this relation is obscured, however, by using the radius of the circle instead of the diameter, producing the formula $C = 2\pi r$. Perhaps the version using the radius is popular because in school and throughout later life you also encounter the formula $A = \pi r^2$ for the area of a circle, which is much simpler with r than with d—with d it becomes $A = \pi d^2/4$. But people remember the area formula in the "pie are squared" version much more than other formulas involving circles. Even my wife once bought far too much fringe for a circular table-cloth because she calculated πr^2 when she needed πd.

The Greek letter π corresponds to "p" in the Roman alphabet. It was chosen by English mathematicians about 300 years ago for its mathematical purpose because it is the first letter of two Greek words for circumference, from which we derive "perimeter" and "periphery." We learn to use the rational number 22/7 for π, but π itself is not a rational number, that is, it cannot be represented exactly by a fraction whose numerator and denominator are integers Lambert's proof of the irrationality of π did not come until 1761. Like all irrational numbers, π does not have a repeating pattern of any kind when it is expressed as an infinite decimal, unlike such rational numbers as 2, which is 2.0000 . . . or 1/3, which is 0.333 . . . or 5⅐, which is 5.142857 142857 142857 142857 . . . .

Even more, as proved by Ferdinand Lindemann in the second half of the nineteenth century, π is a transcendental number, that is, it is not the solution to any algebraic equation where the coefficients are rational numbers (*see* **Transcendental Family,** pp. 290–292).

SIMILAR SPECIES Although π is irrational—and even weirder than that in some ways—it has a long history of rational approximations that go far into the past. King Solomon's Temple, as described at 1 Kings 7, included a "molten sea" that was "ten cubits from the one brim to the other: it was round all about . . . a line of thirty cubits did compass it round about." Legend has it that fundamentalist politicians in various American states have embodied the

"Biblical value" of $\pi=3$ into state laws, although the actual laws are not found on closer observation.

Egyptian mathematicians gave a somewhat more accurate value than had the Hebrew writers of the Bible, one that in modern terms has sometimes been translated as 3.1605. This result comes from working backward from problems that use as a formula for the area of a circle $A = (8/9d)^2$. If you compare this value, which expands to $64d^2/81$, with the modern formula for the area in terms of the diameter, which is $\pi d^2/4$, and solve for π, the result is 256/81, from which the approximation 3.160494 can be derived.

In the third century B.C. Archimedes showed that π is less than $3\frac{1}{7}$ and greater than $3\frac{10}{71}$, using a convincing argument based on the perimeters of circumscribed and inscribed polygons that bracket the circumference of the circle. As a result of this feat, π is sometimes called Archimedes' Constant. He continued the process through a polygon of ninety-six sides. His result was improved upon a hundred years later by the Alexandrian astronomer Claudius Ptolemy, who worked with the chords of circles (essentially twice the sines of the central angles) and produced a value for π that would translate in modern notation to 377/120, or 3.14166.

About A.D. 460 the Chinese astronomer and engineer Tsu Ch'ung-Chih, working with his father Tsu Keng-Chih, developed an astonishingly close estimate. One account has the pair measuring a large circle about 10 meters in diameter to determine that π is 355/113, which translates to 3.1415929203, very close to the true value (to ten decimal places) of 3.141596536. It seems likely, however, that, like Archimedes, they calculated rather than measured. About a thousand years later, the Arab Al Kashi determined π to be 3.1415926535897932, and an Icelandic manuscript from about A.D. 1300 gives as a general rule that "the circumference of a circle is some three times longer than its breadth and a seventh of the fourth breadth," which means 3½ or 22/7. It is interesting to see how this writer expressed the fraction: clearly he is using the idea of measurement along a continuum, since the "1/7" he refers to is a part of the span from 3 to 4 and not an abstract 1/7.

In Europe, late in the sixteenth century, Valentin Otho and Adriaen Anthoniszoon each rediscovered, almost by accident, the Chinese ratio of 355/113, which each thought was the "true" rational value of π. At about the same time, François Viète [French, 1540–1603], the mathematician who helped shape trigonometry and gave us the notation of algebra as we know it today, made a great advance in the calculation of π, determining that its value was between 3.1415926535 and 3.1415926537. He used Archimedes' method, with the advantage of decimal notation. In 1596 Ludolph van Ceulen [German–Dutch, 1540–1610] calculated π to thirty-five decimal places, basing his final approximation on a polygon of 2^{62} sides inscribed in and circumscribed about a circle. Van Ceulen, who worked on π almost all his life, requested that the thirty-six digits of π be inscribed as a fitting epitaph on his tombstone. This was apparently done, but efforts by to find the tombstone have failed. In memory of his achievement, Germans, and the occasional English-speaking pedant, sometimes call π the Ludolphine number.

The number π reached maturity with the independent invention of the calculus by the English mathematician and physicist Sir Isaac Newton [1642–1727] and the German Leibniz. The method of circumscribed and inscribed polygons was abandoned and various limits based on infinite calculations came into vogue. John Wallis contributed one of the most famous infinite products:

$$\frac{\pi}{2} = \frac{2}{1} \times \frac{2}{3} \times \frac{4}{3} \times \frac{4}{5} \times \frac{6}{5} \times \frac{6}{7} \times \frac{8}{7} \times \frac{8}{9} \times \cdots$$

The following infinite series or improved variations on it became the tool of choice for calculating decimal approximations of π:

$$\frac{\pi}{4} = \frac{1}{1} - \frac{1}{3} + \frac{1}{5} - \frac{1}{7} + \frac{1}{9} - \frac{1}{11} + \frac{1}{13} - \frac{1}{15} + \cdots$$

This series is less mysterious when interpreted in the light of calculus. A relatively simple fraction when integrated (one of the fundamental operations of calculus) gives values that are angles measured in radians (a radian is the angle formed when two radii of circle cut an arc equal to the length of the radius). Such an integral can also be represented as an infinite sum. Common angles when measured in radians are expressed fractions of π. For example, $\pi/4$ is the radian measure of an angle whose degree measure is 45°. Thus, this series is another way of expressing the value of the angle 45°.

Although having a good value for π to about ten decimal places is useful for measurements in astronomy or precision machining, no one needs π to 35 or 100 places or more. Yet throughout the ages and even still today people obsessively calculate more and more places. Using a version of the series above for the angle $\pi/6$ (= 30°), Abraham Sharp, in 1699, calculated π to 71 decimal places. Two other early calculations based on this series were the 1719 calculation of a French mathematician that reached 112 decimal places correctly and an English calculation in 1841 that reached 208 decimal places, although later it was shown that only the first 152 were correct. The same calculator returned to the task in 1853, picked up where he went wrong, and then reached 400 places correctly. Meanwhile, Zacharias Dase, a lightning calculator employed by Gauss, worked out 200 places in 1844 using the version for $\pi/4$ given above. Finally, in 1873, William Shanks, an English mathematician, achieved a curious kind of immortality by determining π to 707 decimal places, which everyone took to be definitive and the ultimate in calculating perseverance. It took Shanks some twenty years to get that far.

Good calculating machines became common over the next hundred years. After the trauma of the Great Depression and World War II, several workers turned to recalculating the decimal value and soon discovered that Shanks had made an error at the 528th place, and all the digits beyond that are wrong. About this time, however, digital computers were invented and the history of calculating π soon changed from heroic personal efforts to longer and longer runs of electronic calculation. In 1949 the electronic computer ENIAC was used for seventy machine hours to calculate π to 2037 decimal places; five years later a different computer carried π to more than 3089 decimals, taking thirteen minutes. In England a smaller computer calculated π to 10,000 decimal places in 1958, but it was discovered that, like Shanks, the machine had erred. The computer miscalculated the 7480th decimal place and therefore all subsequent digits. Computers continued to be used in pursuit of π, and in 1961 two scientists using an IBM 7090 computer took nearly nine hours to determine the value to 100,625 decimal places.

In 1999 a computation of π by two Japanese mathematicians, Yasumasa Kanada and Daisuke Takahashi, carried π to 206.15843 billion decimal digits. Their original calculation took more than thirty-seven hours, and then they spent another forty-six hours verifying the digits, not only by running the calculation again, but also by using a different method entirely to reach the same set of digits. Earlier they found eight places where the sequence 0123456789 appears and eight with the sequence 9876543210. Starting at the 12,479,021,132th decimal place there is a sequence of a dozen 9s in a row. One of the oddest sequences of digits found was 27182818284, which started at the 45,111,908,393th place—you perhaps recognize the digits as the first part of e.

The computers no longer have to rely on the old infinite series and products. Since 1995 scientists have been able to use a formula to find any given digit of π provided π is expressed as a binary numeral, that is, as a string of 1s and 0s. In binary form the first few digits of π are 11.0010010000 instead of 3.141592653, but the binary digits, like the decimal ones, cannot repeat in any pattern because of π's irrationality. The formula for the nth digit of the decimal expansion of π has been used to find the five-trillionth digit, which is 0, but that is not much use since the only way to find the ones before it would be to calculate them one by one (or 1 by 0 as the case may be).

Although hard to resist, the pursuit of more and more digits for π is mostly a useless avocation. The nineteenth-century astronomer and mathematician Simon Newcomb [Canadian-American, 1835–1909] once remarked of π, "Ten decimal places are sufficient to give the circumference of the whole visible universe to a quantity imperceptible with the most powerful telescope."

One paradoxical use for generating long strings of digits for π is to use the digits as random numbers. Although π passes all statistical tests for randomness, it is clearly not a random number, since each digit is completely determined in advance. The randomness of π's digits is somewhat disconcerting to those who feel that the simplest of curves should have a less disheveled ratio between the way around and the way across, but randomness is actually very difficult to define in any meaningful way. So even if the digits appear to be random, we know they are not, and a greater intelligence than ours would see this instantly.

PERSONALITY   Since most people find it easier to memorize words than a long series of numbers, several sentences or verses have been suggested as a means of remembering at least the beginning of the sequence of digits in the decimal expansion of π. For example, the short phrase "Yes, I have a number" has the number of letters in each word that corresponds to the value of π to four decimal places (3.1416). Another version using the same method is "May I have a large container of coffee?" which brings the approximation to 3.1415926. James Jeans, perhaps in reaction to this caffeine-laden mnemonic, is the creator of the best-known memory aid: "How I want a drink, alcoholic of course, after the heavy chapters involving quantum mechanics" (3.14159265358979).

ASSOCIATIONS   No matter how the subject of mathematics is approached, π forms an integral part. It is mere coincidence, a mere accident, that π is defined as the ratio of the circumference of a circle to its diameter. If mathematics had progressed on a different path, so that geometry did not become important early, but arithmetic and probability were its basis, then π would have emerged long before anyone noticed that it was connected in some way to the circle. When geometry at last became a part of mathematics, then mathematicians would have been astonished to discover that π turns up in the formulas for the circumference and area.

Radian measure, which is the natural measurement system for angles, employs π as part of the familiar right angle (in degree measure 90°) and straight angle (180°). As in all systems for measuring angles, including degree measure, radian measure is based upon using arcs of a circle to assign numbers to angles (think of how a protractor works). Instead of dividing a circle into 360° (thought to have come from Mesopotamian astronomy) the radius of the circle in radian measure is called 1 and laid out along an arc of the circle. The central angle of the circle that cuts off that amount of arc is an angle whose measure is 1—that is, 1 radian, but since the radian system is natural, neither the word *radian* nor any special symbol, such as °, need be employed.

Since the radius of the circle is taken to be 1, the circumference, from the formula $C = 2\pi r$, is 2π. The straight angle, which can be taken as intercepting a semicircle, is π, while the central angle that intercepts a quarter circle—that is, the right angle—measures $\pi/2$. When necessary to find an equivalence between degrees and radians, it is convenient to use the relationship $\pi = 180°$.

Radian measure is a third model, after the number line and infinite decimals, for the real numbers. Expressed in radians, the size of an angle is as much a number as the point on the number line. This works because the angle is pictured as a rotation of one side about the vertex while the other remains stationary. Angles measured this way can be positive (measured counterclockwise) or negative (measured clockwise) and, although 2π, or -2π, takes the rotating side completely around the circle, there is no reason to stop there and angles corresponding to any real number can be formed. A number such as $\sqrt{2}$, for example, is a bit more than a right angle, while a real number such as -100 requires rotating the side of the angle clockwise for nearly sixteen trips about the vertex.

Augustus De Morgan [English, born in India; 1806–1871], the first Professor of Mathematics at University College,

London, and the first president of the London Mathematics Society, worked primarily in developing the early ideas of symbolic logic. He also collected mathematical oddities and anedotes, published as *Budget of Paradoxes*. One concerning π comes from his own life. In conversation with an actuary about the chances that at the end of a given time a certain proportion of a group of people would be alive, De Morgan quoted the formula for this calculation, which involves π. De Morgan interpolated into the discussion the geometric meaning of π as the ratio of the circumference of a circle to its diameter. The actuary was astounded at this revelation: "My dear friend," he exclaimed, "that must be a delusion. What can a circle have to do with the number of people alive at the end of a given time?"

Here is a problem that combines probability theory with number theory. If two numbers are picked at random from the set of integers in the decimal expansion of π, what is the probability that they will have no common divisor? The surprising answer is $6/(\pi^2)$. This is actually an application of the idea that the sequence of digits in π is in some sense a random sequence, for the same result will occur for any natural numbers chosen at random, not just the digits in π.

One of the most curious methods for computing π is attributed to the French naturalist and mathematician Georges-Louis Leclerc, Count Buffon [1707–1788]. A flat surface is ruled by parallel lines that are all the same distance apart. A needle of length equal to the distance between the lines is dropped on the ruled surface. If the needle lands so that it crosses a line, the toss is considered favorable. In a long run of tosses, the ratio of successes to failures—that is, the probability of a favorable toss—is $2/\pi$. This surprising idea has been tested many times by tossing needles in reality or by simulating tosses with a computer program. The ratio converges toward a value that provides π correct to several decimal places after only a few thousand tosses.

Define the following function for natural numbers n: The value of $f(n)$ for any n can be found as the end result of a finite sequence that begins by subtracting 1 from n and then find the first multiple of $(n - 1)$ that is greater than or equal to the original number; and repeat the process beginning with the $n - 2$ but look for a multiple greater than or equal to the previous multiple. For example, if $n = 8$, then $n - 1 = 7$, and the first multiple of 7 greater than or equal to 8 is 14. Start again with $8 - 2 = 6$; the new multiple will be 18, which is the multiple of 6 greater than or equal to 14. Then start with 5 and find 20 as the multiple of 5 greater than or equal to 18. The multiple of 4 will be 20 also, since it is equal to the previous result. The multiple of 3 will be 21, and of 2 will be 22, which of course is also the multiple of 1, but who's counting? In any case, the value of $f(8)$ is 22 according to this method. Similar reasoning shows that $f(2) = 2$, $f(3) = 4$, and $f(10) = 34$. Clearly this function is defined for all natural numbers greater than 1 and is generally increasing.

Where does π come into the picture? The ratio $n^2/f(n)$ goes to π as n becomes larger and larger.

Genus *Complex* (All-Inclusive Numbers)

Like most people who come to love mathematics, I was infatuated at an early age, so I read popular accounts and histories of mathematics in my early teens, well before I had any formal mathematics training beyond computation, measurement, and recognition of geometric shapes. In my reading I found fleeting references to numbers beyond the real numbers. These numbers were based upon what seemed an extremely mysterious idea, that there must exist an entity which, when squared, would produce the number -1. I failed to see how this could be, since I already understood that the square of every real number is positive. The name assigned to numbers with negative squares only reinforced my feeling that such entities did not really exist, because they were called *imaginary numbers*, a term introduced by Descartes. Descartes needed the imaginaries to regularize a rule about the solutions of equations but did not think they had any real purpose. I equated "imaginary" with "nonexistent."

Thus, I was unprepared to read that discoveries about the imaginary numbers and the complex numbers, which combine imaginary with real numbers, are considered important advances in mathematical thought, although not always recognized as such at the time. The Renaissance mathematician and physician Girolamo Cardano [Italian, 1501–1576] observed that although he could find complex-number solutions to simple equations, the solutions were sophistic, subtle, and useless. Well before Cardano, in the first century A.D., the inventor and scientist Hero (or Heron) of Alexandria had contemplated complex numbers but had dismissed them. Complex numbers were sometimes recognized by later mathematicians, with Leibniz making the odd remark that they were halfway between existing and not existing, rather like the Holy Ghost.

An equation involving an imaginary number that is as mysterious as it is fascinating was first published by Euler as early as 1748. Often this equation is described as showing how the five most important numbers in mathematics are related. The numbers are the base of the natural logarithms, symbolized by e; the ratio of the circumference of every circle to its diameter, symbolized by π; the imaginary square root of -1, symbolized by i; the identity element for multiplication, or 1; and the identity element for addition, or 0. The equation, known as Euler's Formula, is

$$e^{\pi i} + 1 = 0$$

The American mathematician Benjamin Peirce [1809–1880], after rediscovering this equation on his own, is said to have walked into his classroom one day and derived the equation for his students. He told the class: "Gentlemen, that is surely true, it is absolutely paradoxical; we cannot understand it, and we don't know what it means, but we have proved it, and therefore, we know it must be the truth." Peirce overstated the paradoxical nature, however. Once one becomes familiar with the general background, it is easy to see that $e^{\pi i} + 1 = 0$ is a perfectly logical consequence of one of the great applications of complex numbers to trigonometry.

The mathematician Abraham De Moivre [Franco-English, 1667–1754] is thought to have been the first, in 1722, to recognize the important relationship that is the foundation of the trigonometric interpretation of complex numbers, although he did not express it in the form that is known today as De Moivre's Theorem. The great eighteenth-century mathematician Leonhard Euler mined the same ideas as De Moivre somewhat more extensively and also found that the trigonometric functions could be expressed in terms of complex powers of the number e. From Euler's work to the equation that puzzled Peirce is a baby step for anyone with the slightest knowledge of trigonometry. (These relationships are explored further at $1 + i$ and e^i.)

In more advanced mathematics, the role of complex numbers in mathematics begins to be more obvious in other ways. Nearly every theory is simpler in application if complex numbers are included. Just as Descartes found his rule for solutions to equations easier to state when complex numbers are allowed, so are many other rules, including the one known as the fundamental theorem of algebra. In calculus, integration is often possible only if complex numbers are permitted, although one normally discards the imaginary solutions in applying the results. The theory of functions of a complex number is in many ways simpler than the theory of functions of a real number. Difficult problems in fluid flow can be solved when the flow is expressed in complex numbers because of the simpler behavior of complex functions. The "fluid" may be heat or electric current, two of the most important applications of complex numbers.

The common way of showing a complex number is $a + bi$ where a and b are both real numbers—a is called the real part of the complex number and the real number b is called the imaginary part (not bi as you might expect). With this notation and the ordinary meanings of the operations of addition and multiplication, all mathematics can be extended to cover operations with complex numbers (although it is far from obvious how one will treat extracting the roots of any complex number). If c and d are also real numbers, then $c + di$ is also a complex number. The sum $(a + bi) + (c + di)$ is found by adding the two real parts, a and c, and then adding the two imaginary parts, b and d. The result is $(a + bi) + (c + di) = (a + c) + (b + d)i$. This is clearly a complex number itself, so the complex numbers are closed for addition. Similarly, a little more algebra demonstrates that $(a + bi) \times (c + di) = ac + adi + cbi + bd(i^2)$, but since $i^2 = -1$, this can be rewritten as the complex number $(ac - bd) + (ad + bc)i$. Thus, the complex numbers are closed for multiplication as well.

It still seems amazing to me that the use of complex functions has been one of the great forward movements in the study of the natural numbers. The progress began in 1859 when Bernhard Riemann [German, 1826–1866] used complex functions to investigate a long-standing conjecture (originally made by Gauss) that the number of prime natural numbers less than N (which is hard to determine by counting if N is large) is approximately the same as the ratio of N to ln N (*see* discussion of *ln, the natural logarithm,* at **Genus Real, 2.71828 . . .**). In the process, Riemann developed a function, now famous to mathematicians, called the zeta function, or $\zeta(z)$, which is a limit of an infinite series (like some of the series used to calculate π). Here is the series, where z, as is common in mathematics, represents a complex number:

$$\zeta(z) = \frac{1}{1^z} + \frac{1}{2^z} + \frac{1}{3^z} + \cdots$$

The zeta function is best known for a conjecture that Riemann made about it that, if true, can be used to establish many interesting results, but for our purposes it is most appropriate to note that this function was used near the end of the nineteenth century by mathematicians Jacques Hadamard [French, 1865–1963] and Charles J. de la Valée Poussin [Belgian, 1886–1962] to show that the conjecture about the number of primes is indeed true. This is only one example of how complex functions have come to have a role throughout number theory and, in fact, throughout mathematics.

Three Sometimes-Seen Species

$i \quad \sqrt{-1}$ The Imaginary Unit

FIELD MARKS In mathematics today, the word "imaginary" is viewed as something of an embarrassing hangover from past prejudices, for $i = \sqrt{-1}$ exists in whatever Platonic ideal world that all the other numbers do. Nevertheless, the terminology sticks, and so we call i the imaginary unit and the products of real numbers with i are known as the family of imaginary numbers. In my view, which is based more on intuition than on absolutely compelling evidence, i is much more acceptable than $\sqrt{2}$; but there are only a few of us unreconstructed Pythagoreans left who are uncomfortable with irrational numbers.

In one way, however, it is more difficult to accept i's existence today than it was when the number system was less well understood. Early mathematicians often found that their equations or even some apparently practical problems had among their solutions either negative numbers or numbers that were the square roots of negative numbers. Both kinds of number were viewed as equally nonexistent and useless. Today, however, the representation of numbers as equally spaced points on a line is one of the main models for numbers. On a line, it is easy to see where the negative numbers can be placed. It is also easy to see that when one has filled up the line with all the real numbers, including the Irrational Family, there is no place left to fill. Thus, the square roots of negative numbers seem beyond understanding at first. Even after i had been used sparingly for over a hundred years, Euler wrote (in 1768) that "because all conceivable numbers are either greater than 0 or less than 0 or equal to 0, then it is clear that the square roots of negative numbers cannot be included among the possible numbers. Consequently, we must say that these are impossible numbers."

Euler failed to anticipate the solution that makes the numbers not only possible but even sensible. Putting the mysterious square roots of negative numbers on a line *perpendicular* to the real number line, however, even though it eventually becomes familiar, may seem at first like an artificial solution to handling otherwise impossible numbers.

SIMILAR SPECIES Although i is the prototype square root of a negative number, the same considerations that suggest assigning a number to be the solution to the equation $x^2 = -1$ also apply to $x^2 = -2$ or $x^2 = -4$ or to any equation of the form $x^2 = -k$ where k is some positive real number. Thus, we could describe a whole family of imaginary numbers as numbers of the form

$$\sqrt{-x^2}$$

where x is a real number. Note that x^2 is always positive; so $-x^2$ is always negative. In practice, however, it works out better to represent the square root of $-x^2$ as the number ix, which means $i \times x$ and nothing fancy. Notice that $(ix)^2 = i^2 \times x^2 = -x^2$. Furthermore, this notation makes it easy to represent

$$-\sqrt{-x^2}$$

as $-ix$.

In some ways i is like 1 and in others it is like -1.

Like 1, i is the unit from which an entire infinity is built (*see* discussion at **Kingdom Infinity, c**), since the infinity of the imaginary numbers is easily seen to be of the same size as the real numbers and the correspondence is as simple as it could be. If x is a real number, then the corresponding imaginary is ix and vice versa.

On the other hand, i is also an operator, just as -1 is. When -1 is an operator, it changes a number into its opposite; the operator -1 is the same as multiplying by -1. Operate with -1 twice and the number is returned to its original value. When the operation is i, it changes a real number to an imaginary one and an imaginary number to a real one. But unlike -1, two operations with i do not bring you back to the starting place. Consider starting with 2. The first operation produces $2i$, but a second operation on $2i$ results in $2 \times i \times i = 2 \times -1 = -2$. A third produces $-2i$, while a fourth will get you back to 2. Mathematically, multiplication by -1 is like a 180° turn, while multiplication by i is like a 90° turn. Some treatments of complex numbers use the rotation property of i as the basis for the whole approach to imaginary numbers.

The vector i (note the boldface type, conventionally used for vectors, which have both magnitude and direction and are drawn with arrows), is another entity closely related to i. On a three-dimensional graph, which includes a z axis that projects from the plane of the page, i represents a unit vector along the positive x axis; it is usually teamed with j and k, unit vectors along the negative y axis and the z axis. But j, along the y axis, is more like i than i is.

PERSONALITY   One virtue of thinking of i as a rotation of $90°$ is that it makes it somewhat easier to understand how i is related to $-i$. The defining characteristic of i is $i^2 = -1$, but

$$(-i)^2 = [(-1)(i)]^2 = (-1)^2 \times i^2 = 1 \times (-1) = -1$$

also; so the thoughtful reader says "Wait a minute! If $\sqrt{-1} = i$, how does it also equal $-i$?" Here one of the conventions of mathematics saves the day. Recall that although $1^2 = 1$ and $(-1)^2 = 1$, we agree that $\sqrt{1} = 1$ and that we solve the equation $x^2 = 1$ as $x = \pm 1$. So it is with i: we agree that $\sqrt{-1} = i$ and that the solution to $x^2 = -1$ is $x = \pm i$.

Euler tried unsuccessfully to explain how to compute with imaginary numbers. One notable example, in his *Algebra*, was his observation that $i\sqrt{-4}$ is equal to 2, since it must be the same as

$$\sqrt{(-1)(-4)} = \sqrt{4} = 2$$

This reasoning may be flawless, but it does not produce correct and consistent answers in problems, which is the ultimate test. Computation with the square roots of negative numbers is to be avoided. Instead, all square roots of negative numbers should immediately be converted to a form using i. Thus, instead of calculating $i\sqrt{-4}$, Euler should have written the problem as $i \times 2i$, from which he would have easily seen that the answer is $2 \times i^2 = -2$ and not 2.

ASSOCIATIONS Although everyone else uses i for $\sqrt{-1}$, electrical engineers and some physicists use j because they have already assigned i to mean "current." It is possible to explain alternating current without using $j = \sqrt{-1}$, but the need to deal with vectors and trigonometry at each stage becomes rather complicated. Treating the equations for current in terms of imaginary numbers greatly simplifies both the computation and the interpretation. It has been said that the fundamental theory of radio would be impossible to understand without i and that many electrical devices would never have been developed if it had not been for the theory of electricity based on i.

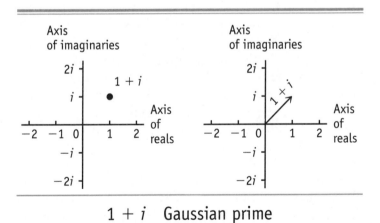

$1 + i$ Gaussian prime

FIELD MARKS The number $1 + i$ is chosen for discussion in this field guide as a representative complex number, not because it has any particular magic of its own.

Imaginary numbers are interesting but on their own not especially useful. The main difficulty is that as a set they are not closed for multiplication, since the product of two imaginary numbers is always a real number. Moreover, real numbers, which are closed for multiplication, do not lead to the imaginaries through either of the two basic operations of addition and multiplication, although one can get from the real numbers to the imaginaries by the derived operation of extracting a root. Thus the bridge between the real numbers and the imaginaries is very narrow and only one way. But as soon as you add a real number to an imaginary, you enter a new realm in which the narrow one-way bridge is replaced by a superhighway. Any possible operation with numbers can be described and enacted using the sums of real numbers and imaginaries, which form the set known as complex numbers.

SIMILAR SPECIES One of the reasons that complex functions are so useful is that each complex number is usually pictured as a pair with a real part and an imaginary part—technically, the "imaginary part" is the real multiplier of i and not itself an imaginary number. This pair of numbers can be shown on a plane exactly the same way that an ordinary point represented by two real coordinates is shown on the plane (*see* the illustration on page 311 and also the discussions of using ordered pairs to represent integers at **Genus Integral** and to represent rational numbers at **Genus Rational**). Such a representation has the advantage that the complex numbers can be seen to be as logically correct as the real numbers since the difficult imaginaries are replaced with real numbers and an unusual operating rule for multiplication. Indeed, as I pass mentally from the natural numbers that Leopold Kronecker viewed as ordained by God to the complex numbers, the stumbling block, if any, is between the rational numbers, easily seen as pairs of integers that are in turn pairs of natural numbers, and the real numbers, which are not *pairs* of rational numbers but most often defined as infinite sets or infinite sequences of rational numbers. Then the final step from real to complex returns to the simple concept of the numbers as pairs.

The definition of complex numbers $x + iy$ where x and y are both real numbers and $i = \sqrt{-1}$ is changed easily into a definition for real-number pairs (x, y). When placed on a coordinate plane, these pairs appear as points exactly where the ordinary real-plane points would be, but now each point indicates a single complex number. The set of points on a plane, when they are interpreted this way, form the complex plane. A representation of the complex plane is known as an Argand diagram, after the Swiss accountant and amateur mathematician, Jean Robert Argand [1768–1822]. It is the same as the real plane, but the vertical axis is iy. We can define two complex numbers as equal if they map to the same point, which shows that there are exactly as many complex numbers as points on the plane. This relationship of the complex numbers to the plane is as definite and as useful as the relationship of the real numbers to the line.

One of the insights that comes from observing the complex numbers as points on a plane is that the common relations "is greater than" and "is less than" do not apply to complex numbers. There is a related idea, however. We define the absolute value of a real number as its distance from 0 on the real line. Similarly, the absolute value of a complex number is its distance from 0 on the complex plane. Even though one complex number is neither less than nor greater than another, its absolute value (a real number) does obey these ordering relationships. Notice that the Pythagorean theorem can be used to define the absolute value of a complex number, since the distance is the hypotenuse of the right triangle that has x and y as its sides. Thus, for the complex number (x, y) the absolute value is simply the square root of $x^2 + y^2$.

One might wonder if prime numbers exist for complex numbers. Karl Friedrich Gauss cleverly investigated this topic and concluded not only that the answer is yes, but also that there is an equivalent to the fundamental theorem of arithmetic (that every natural number can be factored into primes in only one way) for the complex numbers. First, one needs to define a complex integer. The most obvious definition is simply that a complex integer is a complex number of the form $a + bi$ where a and b are integers. These complex numbers are known as the Gaussian integers. Gauss also worked out a definition for primes such that every nonzero Gaussian integer can be factored into primes and powers of i in one and only one way, provided the primes are in the group that Gauss defined as "positive." Gauss's rather complicated system produces a rule that for a Gaussian integer to be "positive" the real part must be greater than or equal to the absolute value of the imaginary part—so, for example, $1 + i$ and $3 - 2i$ are positive, but $1 + 2i$ and $3 - 6i$ are not.

For the record, the first few "positive" Gaussian primes are $1 + i$, $2 + i$, 3, $3 + 2i$, $3 - 2i$, $4 + i$, $5 + 2i$, $5 - 2i$, $6 + i$, $6 - i$, $5 + 4i$, $5 - 4i$, 7, and $7 + 2i$. Notice that not all prime integers from genus *Integer* are also Gaussian primes. In the list so far, you find 3 and 7, but not 2 and 5. Indeed, $2 = (1 + i)(1 - i)$ and therefore is not prime and 5 is $(2 + i)(2 - i)$. A general rule states that the genus *Integer* primes larger than 2 are also Gaussian primes if and only if they cannot be expressed as 1 more than a multiple of 4. Thus, $13 = 4 \times 3 + 1$ and $17 = 4 \times 4 + 1$ are not Gaussian primes, but 19 and 23 are.

PERSONALITY Operations with complex numbers are easily understood in terms of the ordinary algebraic addition and multiplication algorithms for binomials, except that i^2 must be replaced with -1. For example, to add the real-number binomials $2x + 3$ and $4x - 5$, you simply add the x terms and the constant terms separately to get $6x - 2$. Similarly, for complex numbers $2 + 3i$ and $4 - 5i$ the sum is $(2 + 4) + (3 - 5)i = 6 - 2i$. Multiplication follows the FOIL method for real binomials. The product, such as for binomials $(2x + 3)(4x - 5)$, is the sum of the products of the

First terms (in this case $2x \times 4x = 8x^2$),

Outer terms (in this case $2x \times -5 = -10x$),

Inner terms (in this case $+3 \times 4x = 12x$),

Last terms (in this case $+3 \times -5 = -15$),

which is $8x^2 + 2x - 15$. Using the same method for $(2 + 3i)(4 - 5i)$ gives $8 - 10i + 12i - 15i^2 = 8 + 2i + 15 = 23 + 2i$.

These rules that derive from real-number algebra can then be translated into the form of rules for complex numbers as ordered pairs of real numbers. Two complex numbers (a, b) and (c, d) are equal if for the real numbers $a = c$ and $b = d$. The sum of (a, b) and (c, d) is simply the pair $(a + c, b + d)$. The product is a little trickier, but easily derived from the FOIL rule, since the first member of the ordered pair for the product is the algebraic sum of F and $-$L, while the second member is the algebraic sum of O and I: that is, the product $(a, b) \times (c, d)$ is $(ac - bd, ad + bc)$. For the most part, for numbers at least, if you know how to define equality, addition, and multiplication, everything else follows.

With complex numbers, every real number has more roots than the obvious real one or two. Consider 64, for example. We are familiar with the two square roots, 8 and -8. If we stick to real numbers there is only one cube root, 4, since $4^3 = 64$. You know that $(-4)^3 = -64$, so -4 is not a second cube root. Now put complex numbers into the picture. There are several ways to show that 64 also has the two complex cube roots $-2 + 2i\sqrt{3}$ and $-2 - 2i\sqrt{3}$ (we will look more closely at one of these below) as well as 4, making three cube roots in all. Then we might suspect that there will be four complex fourth roots of 64, including the real ones $+2\sqrt{2}$ and $-2\sqrt{2}$. In this case, it is easy to determine that the complex roots are $+2i\sqrt{2}$ and $-2i\sqrt{2}$, since $i^4 = +1$.

One feature of complex numbers that is a surprise to people used to real numbers is that the complex numbers are not ordered. As Euler remarked in arguing against the existence of complex numbers, the inequality relationship for complex numbers does not apply. It is impossible to say whether or not $1 + i$ is greater than or less than i, for example, although it is clear that the two complex numbers are not equal.

Since the complex numbers are not ordered, there is no least complex number for an arbitrary set of them. Thus, the "proof" that every number is interesting, which depends on the least member of the set of uninteresting numbers being, on that account alone, interesting, also fails to apply. Perhaps that is why most complex numbers seem rather dull, no matter how useful they are individually and as a class.

As noted above, the notion of order for complex numbers can be replaced to some degree with the idea that each complex number (x, y) has an absolute value that is

$$\sqrt{x^2 + y^2}$$

If we use the common convention of letting z stand for a complex number, then if $z = (x, y)$, the absolute value is

$$|z| = \sqrt{x^2 + y^2}$$

The absolute value of a complex number is sometimes written with pairs of vertical lines as $\|z\|$ and also may be called either the *magnitude* or the *modulus* of the complex number. But it is all the same idea, no matter how it is written or what it is called. Personally, I like the parallelism with real numbers, so I prefer single vertical lines and the name "absolute value of the complex number." But other interests often transcend mine, and so perhaps the most common versions use the double lines, as in $\|z\|$, and the name "modulus."

The symbol $|z|$ represents the distance of the complex number z from the origin of the complex plane, just as the absolute value of a real number is the distance from the origin of the real line. Using an arrow (vector) from the origin to point z suggests a different form for writing a complex number, one that has many practical and theoretical advantages. Consider the complex number z as the ordered pair (x, y). The real number x is the same as $|z|$ times the cosine of the angle between the positive horizontal axis and the line or arrow to z (the angle measured in radians, which means that it is just a real number—*see* discussion at **Genus Real,** 3.14159 . . .). Similarly, the real number y is $|z|$ times the sine of the same angle. If we call the angle by the Greek letter theta, θ, the common choice of mathematicians, then $z = (x, y) = x + iy = |z| \cos \theta + i|z| \sin \theta$. This last version with the two trigonometric functions is called the *polar form* of the complex number, which is graphed here:

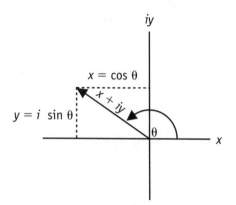

De Moivre's Theorem, alluded to on pages 304–305, is

$$(\cos \theta + i \sin \theta)^x = \cos x\theta + i \sin x\theta$$

This formulation shows its connection to the polar form immediately. If you study this relationship in terms of the previous discussion, then it appears De Moivre's Theorem states that the xth power of any complex number of absolute value 1 has a value that also has an absolute value of 1 and is at an angle that is $1/x$th the angle of the original number. Now think of some specific example to see what this might mean. One of the easiest numbers to contemplate is the complex number 1, which is the same in some ways as the real number 1, but since it is in a different genus must also be a different species. You may prefer to think of this number as $1 + 0i$ or as $(1, 0)$. Since 1 is on the positive x axis, its angle with that axis is 0, and the polar form is $\cos 0 + i \sin 0$—since $\cos 0$ is 1 and $\sin 0$ is 0, this gives back the expected value. In what follows, however, we can take a slightly different angle and get better results. If the angle is taken as $2\pi \ (= 360°)$, the complex number 1 is at the same point, but its polar form is $\cos 2\pi + i \sin 2\pi$.

It is useful to use De Moivre's Theorem when x is set equal to $1/n$, so that it is applied to finding roots rather than ordinary powers. In that case it would show that the 1/3 power (or cube roots) of 1 include a cube root with the value $\cos 2\pi/3 + i \sin 2\pi/3$. (For those still deep into degree measure, the angle is 120°.) But we know that there must be a third cube root of 1. It can be found by starting with 1 in the form $\cos 4\pi + i \sin 4\pi$, which produces via De Moivre's Theorem the third cube root, which is $\cos 4\pi/3 + i \sin 4\pi/3$ (for degree people, 240°).

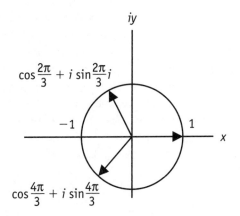

Notice that the points for these are all on a circle of radius 1—the distance from the origin is 1 since $|z| = 1$. Furthermore, they are equally placed about the circumference of the circle. A similar argument shows that the fourth roots, which are the 1/4 power, of 1 are 1, i, -1, and $-i$, equally placed a fourth of the way around the circle. Thus, one of the many applications of De Moivre's remarkable theorem is to explain the nth roots of a complex number and to show what they mean. If we return to the earlier example of the cube roots of 64, the two square roots are on a circle of radius 8 and π (180°) apart, while the three cube roots are seen to be on a circle with radius 4 and $2\pi/3$ (120°) apart. Note that $|-2 + 2i\sqrt{3}|$ is equal to the square root of $(-2)^2 + (2\sqrt{3})^2$, which is the square root of $4 + 12 = \sqrt{16} = 4$. And similarly, the fourth roots of 64 are $+2\sqrt{2}$, $+2i\sqrt{2}$, $-2\sqrt{2}$, and $-2i\sqrt{2}$, spaced around a circle of radius $\sqrt{8}$ at intervals of $2\pi/4 = \pi/2$ (90°).

ASSOCIATIONS A complex function is one that takes every complex number for which it is defined and connects it by some rule to one and only one specific complex number (not necessarily different from the first). The set for which the function is defined can, of course, be shown on a plane; and the set of complex numbers that results from applying the rule can also be shown on the same complex plane or on a different one. So the complex functions provide a useful way to change every point in one plane to points in a second plane. But because the change is effected by complex numbers, there are constraints on how the points from one plane are changed into the other. Often you need to know only a small amount of information about the function to learn a great deal of information about what kinds of changes will happen.

Gauss's construction of the seventeen-sided polygon (*see* discussion at **Genus Natural, 17**) derives in a not very obvious way from De Moivre's Theorem and is, therefore, an application of complex numbers. Gauss showed with this proof how complex numbers could be used to solve problems that seemed far away from $i = \sqrt{-1}$.

Complex numbers have practical applications throughout physics. One example is in the theory known as QED, which in physics doesn't mean *quod erat demonstrandum* ("which was

320

to be demonstrated"), used by geometers of old at the end of a proof, but rather a complex set of rules for computing the behavior of electrons and photons called quantum electrodynamics. In QED the complex numbers are used to derive probabilities of various events that may occur to or in conjunction with subatomic particles. In his 1985 book, *QED*, Richard P. Feynman, whose contributions to QED earned him a share of the Nobel prize in physics in 1965, explains how photons interact with matter first in nonmathematical terms; then, in a footnote, he describes the interactions this way:

> For those of you who have studied mathematics enough to have come to complex numbers, I could have said, "the probability of an event is the absolute square of a complex number. When an event can happen in alternative ways, you add the complex numbers; when it can happen only as a succession of steps, you multiply the complex numbers."

Notice that the absolute square of a complex number such as z is just $|z|^2$.

The only way to satisfy reasonable assumptions about the wave function for the behavior of a subatomic particle, such as an electron, is to make the equation that describes how the wave moves through space into a complex function. More specifically, the Schrödinger wave equation known as Ψ (the Greek letter psi) is a function of real-number locations in time t and space x, but the wave itself is of the form $\cos(kx - \omega t) + i \sin(kx - \omega t)$, where k and ω are constants that are determined by the wavelength and frequency of the wave (which in turn are related to linear momentum and Planck's constant and to total energy and Planck's constant). Thus, the Schrödinger wave equation describes a complex function.

Physicists had a hard time figuring out what this equation means. The i appears because it is a solution for one variable in a system of equations, not because someone had a physical meaning for it. Indeed, one writer on the subject wrote that the i "makes it immediately apparent that we should not give to wave functions a physical existence in the same sense that water waves have a physical existence."

Most quantum physicists came to accept an interpretation of what the complex numbers of the wave equation mean that was developed by the German mathematician and physicist Max Born [1882–1970]. Born started with the notion that the wave function for a particular time and location has a value that is a specific complex number $z = x + iy$. Mathematicians had defined a complex number related to $z = x + iy$, called the complex conjugate of z, or z^*, as $x - iy$. Born argued that the product zz^* for a value of z derived from the Schrödinger wave equation is the probability of locating the particle described by the equation in a given small region of space. Notice that the product zz^* is always a real number, not a complex number, since $zz^* = (x + iy)(x - iy) = x^2 + y^2$, which is the sum of the squares of two real numbers and therefore is itself real. In fact, if the original value is taken to be a complex number z then the probability is seen to be $|z|^2$. Thus, even though the particle always obeys a complex wave function that is difficult to interpret, the function determines a real value that can be checked by experiments. Since the real value is a probability, not a specific location, several trials need to be run to check whether it works or not. When a series of such trials are run, however, every experiment performed so far shows that Born's probability interpretation based on the square of the absolute value is correct—although still a little mysterious.

e^i 0.54030 . . . + (0.84147 . . .)i
cos 1 + i sin 1

FIELD MARKS   At first encounter, it seems impossible to raise a real number, such as e, to an imaginary power. But there are so many instances for which mathematics and engineering students need to perform this unnatural act that it soon loses its initial shock value.

One way to look at e^i is to return to the definition of e as the limit as n increases without bound in the expression $(1 + 1/n)^n$. Another way to say this is to state that $e^1 = (1 + 1/1)^1 + (1 + 1/2)^2 + (1 + 1/3)^3 + \ldots$. This sequence can be generalized for powers of e. Thus, $e^2 = (1 + 2/1)^1 + (1 + 2/2)^2 + (1 + 2/3)^3 + \ldots$ and $e^3 = (1 + 3/1)^1 + (1 + 3/2)^2 + (1 + 3/3)^3 + \ldots$, or, in general, $e^x = (1 + x/1)^1 + (1 + x/2)^2 + (1 + x/3)^3 + \ldots$. Thus, the limit as n increases without bound of $(1 + z/n)^n$ is e^z. The use of z is intended to indicate that this is true for a complex number (which of course includes the case where the complex number is the same as a real number). Thus, one way to envision e^i is as the limit for n increasing without bound of $(1 + i/n)^n$.

It is an even easier computation to use the sequence based on factorials that also defines e^x, but with the same substitution of z for the exponent. Since

$$e^z = 1 + \frac{z}{1!} + \frac{z^2}{2!} + \frac{z^3}{3!} + \cdots$$

for any complex value of z, you need only substitute i for z to obtain a sequence that will converge to e^i.

SIMILAR SPECIES   Euler, who did not even believe in imaginary numbers, used the methods of calculus to show that $e^{i\theta} = \cos\theta + i\sin\theta$. The details are beyond this book, but the equation that Benjamin Peirce is said to have found so mysterious is seen to be a straightforward application of Euler's Formula. If θ is assigned the value π, the formula states that $e^{\pi i}$ is $\cos\pi + i\sin\pi$. Since $\cos\pi = -1$ and $\sin\pi = 0$, Euler's Formula for $\theta = \pi$ yields $e^{\pi i} = -1$, which is somewhat more mysterious when written as $e^{\pi i} + 1 = 0$.

PERSONALITY The applications of Euler's Formula are many. One of the most useful provides the definition of the trigonometric functions in terms of complex numbers. To make this transition, begin by finding that $e^{-i\theta} = \cos(-\theta) + i\sin(-\theta)$. The trigonometric functions for negative angles follow simple rules, although those for cosine and sine differ from each other. The cosine does not change value, so $\cos(-\theta) = \cos\theta$, while the sine of a negative is the negative of the sine, or $\sin(-\theta) = -\sin\theta$. Using these rules for a negative angle, the value of e to the negative $-i\theta$ power becomes $\cos\theta - i\sin\theta$. When you add and subtract this result to Euler's original formula, the result with suitable rearranging becomes

$$\cos\theta = \frac{e^{i\theta} + e^{-i\theta}}{2}$$

and

$$\sin\theta = \frac{e^{i\theta} - e^{-i\theta}}{2i}$$

which are in many ways useful equations.

These equations have inspired an analog of the same idea. The two related functions are known as the hyperbolic cosine (also known as cosh) and hyperbolic sine (or sinh) because they are related to the hyperbola in somewhat the same way as the ordinary cosine and sine are related to a cir-

324

cle. The functions are usually defined with the same equations, omitting the i, so that, for example,

$$\cosh \theta = \frac{e^\theta + e^{-\theta}}{2}$$

with $\sinh \theta$ defined similarly. Instead of θ, a complex variable such as z could also be used. In that case, it is easy to see that $\cos z = \cosh iz$ and $i \sin z = \sinh iz$. The other hyperbolic functions, $\tanh \theta$, $\operatorname{sech} \theta$, $\operatorname{csch} \theta$, and $\coth \theta$, are all defined in terms of $\cosh \theta$ and $\sinh \theta$ in the same way as the usual trigonometric functions are defined in terms of the cosine and sine. For example, $\tanh \theta = \sinh \theta / \cosh \theta$.

ASSOCIATIONS   Any regular wave can be expressed as a multiple of the sine or cosine function, such as $a \sin b(\theta + c)$ where a, b, and c are real numbers, while irregular waves are all versions of the sum of a suitable (usually large) number of multiples of the sine and cosine functions. Amazingly, any smooth and continuous real function (that is, any graph of any ordinary curve) can be expressed as the sum of an infinite number of sine and cosine functions. Even when all the functions involved are real, these waves and curves can often be expressed or understood more easily when the equations have been translated to complex powers of e. The relationship between e^i and waves defined by sines and cosines is the original path to the use of complex numbers for quantum mechanics and for alternating current.

The End of the Line

When mathematicians in the eighteenth century began to use complex numbers more routinely, they considered it likely that if square roots of negative numbers had entered the sacred halls of number, then perhaps fourth roots or sixth roots or even all 2nth roots of negatives would also be new types of numbers. But by 1747 Jean de la Rond D'Alembert [French, 1717–1783] found a proof that this would not happen. His proof was not airtight, but soon other mathematicians filled in the gaps. There were to be no more numbers along this route. Indeed, it gradually became apparent that no more numbers of any kind were to be found.

It is perhaps surprising that the complex numbers are the end of the line for numbers. The one-way bridge from the real numbers to the imaginaries has no counterpart leading from the complex numbers to anything else. Why this is so can be explained in several ways. The solution of equations in one variable leads inexorably from the natural numbers through the rational numbers and integrals to the real and complex (although not to the transcendental). Note also that none of the operations of addition, subtraction, multiplication, division, or extraction of roots—in short, none of the algebraic operations in the solutions of equations—lead to a type of number beyond the complex. But clearly this reasoning is not enough, for the transcendentals emerge both as real and complex numbers but are not obtainable by algebraic means. One way you can get to a transcendental number from algebraic numbers is with infinite sequences, series, or products; these lead to numbers hidden in the points of the number line or the points of a plane that cannot be reached by finite algebra, but infinite sequences of numbers shown as points cannot by their nature lead to numbers not on the complex plane.

So if you want to extend Kingdom Number, you need to travel away from the plane. One can picture, for example, numbers that are to three-dimensional space what complex numbers are to two dimensions. In that case, the plane of complex numbers would be a subset of the space of three-dimensional numbers. But any other potential number system that might have the complex numbers as a subsystem fails to do so because the operations required by such a system, especially the kinds of multiplication that can consistently make sense, fail for such a system.

Consider, for example, the often-encountered example of vectors in three dimensions. Vectors in three dimensions can be represented as arrows that seem very much like the arrow representation of complex numbers in two dimensions. There are several ways of showing such vectors as symbols, two of which are as ordered triplets of the form (a, b, c) where a, b, and c are real numbers and as vector sums of the form $a\mathbf{i} + b\mathbf{j} + c\mathbf{k}$, where i, j, and k are unit vectors along the three axes. It would seem that letting $c = 0$ would result in the same entities as the complex numbers. But triplets of the form $(a, b, 0)$ do not behave for multiplication in either of the two types of vector multiplication with the same results as one obtains for pairs (a, b). Similarly, a little experimentation will show that $a\mathbf{i} + b\mathbf{j} + 0\mathbf{k}$ is not the same entity as $a + bi$. Although vectors are not numbers, it needs to be said that the set of real numbers or the set of complex numbers can each be treated profitably as vectors following the usual rules for one-dimensional or two-dimensional vector spaces.

Although the complex numbers are the end of the Kingdom Number, filling the plane completely and incapable of traveling to higher dimensions, there is another type of entity that is related to numbers rather as plants are to animals. Kingdom Infinity, to which we travel now, is part of the same spiritual world as Kingdom Number, but separate from it in many important ways.

Kingdom Infinity

When it comes to infinity, the mind tends to boggle. It is easy to speak of the infinite extent of the universe or the infinite reach of time, but modern cosmology tells us that the universe, although large, has a boundary and that time certainly has a beginning in the big bang and may have an ending in the collapse of the universe. But if the universe is finite, what is there beyond it? Or what exists before or after time? Thus, even in the real world, questions of infinity persist and perplex.

The same is true of the ideal universe that is mathematics. We casually introduce into geometry lines and planes that are of infinite extent and speak of the infinity of points on a line or in space. We say that the natural numbers are infinite, the rational numbers are infinite, the real numbers are infinite, and the complex numbers are infinite; but we do not often stop to ask whether or not all of the infinities are the same or different. Probably without thinking, most people assume that they are the same. Infinity is infinity, after all—isn't it?

The early Greeks questioned the foundations of the universe as no society had before and possibly more astutely than any since. The problems posed by infinity were not overlooked. Zeno of Elea [c. 490–c. 425 B.C.] examined the questions of whether space and time could be infinitely divided and found that answering either yes or no resulted in paradoxes. By the time of Aristotle, 100 years later, Zeno's paradoxes led to two classes of infinity: an incomplete infinity, meaning that an operation could be continued for as many steps as needed, and a completed infinity that consisted of more objects, actual or ideal, than any very large number. Aristotle firmly rejected the completed infinity, while accepting the incomplete version. A half-century later Euclid carefully worded all his postulates and theorems so as to avoid infinity altogether. Where mathematicians today casually restate Euclid's fifth postulate as "parallel lines do not meet" or say that the number of primes is infinite, Euclid rejected the concept of an infinite line and wrote his postulate in terms of the angle that certain lines make when they meet and proved that there is no greatest prime (which is not the same as an infinite number of primes).

The Greek position on infinity was accepted by medieval scholars (except where God entered into the picture), and the issue was not re-examined until Galileo looked at it in the seventeenth century. He seems to have been the first to notice that two sets of numbers, one with what seems to have many more members than the other (such as the even numbers and all the natural numbers), can still be matched one-to-one. But Galileo's apparent paradox was ignored as mathematicians rushed to invent the calculus, basing it almost entirely upon the ideas of infinity that Zeno had called in question so long ago. The calculus gave the right answers, so few worried about its apparent inconsistencies.

Finally, in the nineteenth century it became apparent that some situations involving infinity were creating difficulties and at times even producing wrong answers. Late in the century mathematicians, notably Georg Cantor, began a careful analysis of what it means to say that a line or space is infinite and especially how this relates to Kingdom Number, where not only are the natural numbers infinite, but between each pair of natural numbers one can find an infinite number of rational numbers, and between each pair of rational numbers and infinite number of rational or real numbers. Furthermore, for every pair of real numbers, there is a complex number, so perhaps the infinity of complex numbers is different from that of the real numbers.

The results—based on Galileo's idea of matching two sets of numbers so that for each number in one set there is exactly one number in the second set and vice versa—were surprising and at first often rejected. There are indeed several kinds of infinity, not just the two discussed by Aristotle. In addition to the potential, or incomplete, infinity that Aristotle accepted, mathematics makes sense only if there are several different completed infinities, one countable and an infinite number (there we go again) of uncountable infinities.

Since this field guide is arranged along taxonomic lines, I am calling the various infinities members of their own kingdom. Infinities can be derived from numbers, but they also have roots in set theory and in the ideas of geometry. If one were to build a taxonomic tree for mathematics like the familiar Tree of Life that shows the evolution of the different kingdoms (or more often of the different phyla) from common ancestors, the Tree of Mathematics as most see it today would have a trunk that is the kingdom of sets and branches showing kingdoms of number, infinity, geometry, functions, and others. But this analogy is not exact, I think, since the kingdom of infinity could have ancestors that include members of all the other kingdoms; so a Tree of Mathematics would be an odd organism that starts with a single trunk, then separates into several different branches that somehow rejoin at the top to form a new entity, Kingdom Infinity.

∞ Increases Without Bound
The Potential Infinite

FIELD MARKS The symbol ∞ is one of the most familiar in mathematics. While you might say in words that the limit of $(1 + 1/n)^n$ is e as n goes to infinity, you would write

$$\lim_{n \to \infty}\left(1 + \frac{1}{n}\right)^n = e$$

Since ∞ is so commonly encountered in mathematics, many people unthinkingly consider it as meaning "infinity," with no qualification. (For the record, the symbol ∞ was introduced by John Wallis in 1655.) It should be clear from the discussion of very large counting numbers that infinity is not a very large number, but the symbol ∞ seems like a number in some basic ways. Although ∞ is read as "infinity" or as "to infinity," a careful look at how ∞ is used and at its meaning suggests that ∞ does not even really belong in Kingdom Infinity, which consists of different types of *completed* infinity. For this reason the symbol ∞ should be read as "without bound." Augustin-Louis Cauchy, faced with ∞, said "only God is infinite." Thus the correct interpretation of the sequence used to define e in the equation above is that n continues to be taken to higher and higher values, which causes the expression to become closer and closer to a specific number (the limit), not that n is replaced by some specific number that can be called "infinity." One of the most popular calculus textbooks in the United States, *Calculus and Analytic Geometry*, by George B. Thomas, Jr., and Ross L. Finney, put it this way: "The symbol ∞ is used to represent 'infinity.' However, we do not use this symbol in arithmetic in the ordinary way. Infinity is not a real number. We used the phrase 'becomes infinite' . . . only as a way to say 'out-grows every preassigned real number.' " Later in the book, discussing the limit as $t \rightarrow 0$ of the function $1/t$, which they indicate is ∞, the authors add, "When speaking of infinity as the limit of a function f, we do not mean that the difference between $[1/t]$ and infinity becomes small, but rather that $[1/t]$ is numerically large" when t is near 0.

Another way to say this is to state that ∞ is a potential or incomplete infinity. A mathematician who says that an expression such as $x > 2$ means that x goes from 2 to infinity is saying that x increases as far as is necessary and, if desired, it can be increased even more. Notice that for the sequence of values for n we are dealing with natural numbers, but the values for x are usually taken to be the real numbers. Indeed, the common way to show $x > 2$ is as an unbounded part of the real number line.

Aristotle made the distinction between a potential infinite and a completed one, claiming that only the potential can exist. He was willing to allow the natural numbers the property of being potentially infinite, but denied potential infinity to geometric figures since an infinite figure would extend beyond the universe. Clearly he thought of lines as real marks, while numbers are Platonic ideas. Aristotle also considered the possibility of subdividing a line or time an infinity of times, leading the way to the concept of the infinitesimal (the potentially infinitely small).

Gauss was very definite on this subject in 1831. He wrote, "I protest against the use of an infinite quantity as an actual entity; this is never allowed in mathematics. The infinite is only a manner of speaking, in which one properly speaks of limits to which certain ratios can come as near as desired, while others are permitted to increase without bound."

Thus, it could be said that ∞ is not an entity at all, but rather a process, although this distinction is not so meaningful in mathematics because even what we think of as processes become physical entities (think of functions, for example). A better distinction might be that ∞ is not abstracted from a set; there is no set that can be described as having ∞ members, nor is ∞ a particular member of a set. As we will see, this makes ∞ different from the other kinds of infinity, each of which is associated with sets in one of these ways.

Generally speaking, modern writings on infinity do not bother with ∞ at all, since ∞ is not like the kinds of infinity studied since the late nineteenth century. Instead, ∞ represents an older idea, one that became somewhat codified with Newton and Leibniz and their followers in the eighteenth century. Despite this apparent neglect, ∞ is the infinity that everyone uses in ordinary mathematics, while the other infinities become important primarily to people working in trying to find a firm foundation for mathematics (a difficult and probably impossible task).

SIMILAR SPECIES In some notations, the symbol −∞ is used to mean the process that might be described as "decreases without bound." If the interval for the solution to a mathematical inequality is all real numbers less than some number, then −∞ can be used to show this, just as ∞ is used to indicate an interval that increases as far as one wishes.

There is another "infinity" that is even more mysterious than ∞. When a limit is taken by letting a function go to 0, there is an explanation, developed in the nineteenth century, of what is happening near the limit that avoids the infinitely small. But when calculus was first developed, the infinitely small was an important part of the whole program. This dependence upon quantities that were treated as vanishing near 0 but not being 0 led to a famous criticism of calculus by Bishop George Berkeley [Irish, 1685–1753] that the whole subject was based upon "neither finite quantities nor quantities infinitely small, nor yet nothing. May we not call them the ghosts of departed quantities?" Berkeley, by the way, thought that the reason that calculus produces right answers from an incorrect, logical basis is that somehow compensating errors are introduced—a point of view oddly like that of the modern physicist explaining why quantum electrodynamics works as a result of differences between $+\infty$ and $-\infty$ (see further discussion below).

Despite Cauchy's view that "only God is infinite," he thought man could define the infinitely small in a reasonable way, which he proceeded to do to his own satisfaction, although not to his successors'. His creation was called an infinitesimal and was a variable that had as its limit 0. Thus, the infinitesimal was not a number, although Cauchy then proceeded to treat it as one, soon getting into trouble. One problem that Cauchy was forced to deal with is that in the integral calculus the basic operation is adding an infinity of infinitesimals to get a specific real number, while the differential calculus is based on the ratio of two infinitesimals having different real-number values in different circumstances. Notice the phrase *differential calculus*. A differential is a mysterious entity; here is a brief account of its bedeviling guises.

In calculus there is a second function related to most common functions that is called the derivative; it is the slope of the first function. One notation for the derivative of a function $f(x)$ is $f'(x)$; while another notation is dy/dx, which is essentially the ratio of two infinitesimals (although that idea is defined away in modern treatments of calculus). Setting these two notations equal to each other produces $f'(x) = dy/dx$, which can be rewritten as $dy = f'(x)dx$. Then dy is called a differential. By looking at the function

$f(x) = x$, for which $f'(x) = 1$, it can be shown that dx is also a differential. If there is a difference between differentials and Cauchy's infinitesimals, it is not apparent here.

In this definition, the differential, like a Cauchy infinitesimal, appears to be a function. Sometimes, however, the differential is defined not on the basis of a function $f(x)$ and its derivative $f'(x)$, but on the value of the derivative for a specific real number. In that case, the differential appears to be more like a single quantity than like a function itself.

Despite the unlikely definitions, the differential remains the basis of calculus and calculus continues to be the foundation of physics and therefore of science in general. One must learn to handle the differential properly not only in both parts of the calculus (it is an indispensable part of the integral), but also in an important branch of analysis called differential equations.

This is not the place to go into the partial differential, another foundation of physics, except to say that despite a name that suggests that it is part of an infinitely small quantity, partial differentials are not difficult to accept once one resolves the problem of the infinitely small in the calculus.

Modern calculus textbooks often contain the warning about differentials: "This has nothing to do with the infinitely small quantities"; they are technically correct. The calculus has been put on a logical foundation that is based entirely on definitions that carefully step around anything that might be infinitely small. Nevertheless, the reason for the warnings should be obvious. Differentials look like the infinitely small, walk like the infinitely small, and even quack like it, although of course they are no such thing . . .

Points and lines at infinity are a necessity in certain kinds of geometry as well as in transformations from one coordinate system to another. The best known points at infinity occur in projective geometry, an important branch of the subject that was developed by Girard Desargues [French, 1591–1661] in 1619 and considerably expanded by him in 1636. In projective geometry, two parallel lines meet at a point at infinity; furthermore, all such points lie on a line that is a line at infinity. Kepler also considered parallel lines as meeting at infinity a few years earlier than Desargues. A similar idea occurs in the theory of transformations, such as the inverse transformation (*see* discussion at **Genus *Integral*, −1**) that maps all points inside a circle onto points outside the circle. In the inverse transformation, a straight line is considered to be a circle with an infinite radius and the point at infinity is the center of such an infinite circle— in other words, the center for a straight line is the point at infinity. In this instance it also is necessary to make two parallel lines intersect at the point at infinity; also, two intersecting lines not only intersect at the ordinary place, but also at the point of infinity.

The projective plane of Desargues is more easily understood in terms of a geometric transformation as well. Just as the inverse transformation carries points from inside a circle to the outside, there are various transformations that carry points on a sphere into a plane (every flat map of Earth is made with one of these projections). A projection that maps the points on a sphere into the plane by means of a line through the center of the sphere results in a plane with properties similar to the projective plane. This is easiest to picture if you think of mapping only the lower hemisphere of a sphere that has its south pole resting on a flat plane into that plane via a line through the center of the sphere. The south pole maps directly onto the point where the sphere rests. A point in, say, Tasmania will map onto the plane at some distance from the sphere. A point in southern India or northern Argentina will map farther away from the sphere.

And a point on the sphere at, say Quito, Ecuador, right on the equator, will map to a point at infinity. The line at infinity is the map of the entire equator. Notice that for a point just south of the equator, the distance from the map of the south pole is very great and that this distance increases as you approach the equator. Thus, the line at infinity is essentially a case of increasing without bound, like ∞, although the difference is that these parts of geometry assume that the line at infinity actually exists; whereas in calculus the statement that something goes to ∞ is not taken to mean that there is an actual ∞ off in the distance.

Although there seems to be no need for a point at infinity at either end of the real-number line, the complex plane is accorded such a point to make it complete. (The existence of this point does not imply "continuing infinitely," as both the real line and all planes do, but is instead a special point like the point at infinity in projective geometry.) In a careful analysis, the complex plane is therefore different from the ordinary coordinate plane, which has no extra point at infinity. The complex point at infinity is usually explained by treating the complex plane as a map of the sphere by projection. In this case, the projection is accomplished by a line from one pole (usually taken to be the north pole) that maps every point onto a plane on which the sphere rests, balanced on its south pole. The map of the north pole is the point at infinity. It may not seem much like a point when viewed on the plane (it seems to be an infinitely receded edge), but from the opposite point of view—mapping the plane onto the sphere—it becomes the point at the north pole.

The point at infinity seems closely related to such ideas from analytic geometry as asymptotes, curves that approach a line continually but never meet it. The tangent curve from trigonometry grows larger and larger as the angle approaches π/2 (90° in degree measure), but is said to be "undefined" in modern trigonometry books. Some older books, however, give the value of the tangent of π/2 as ∞.

PERSONALITY   Although older mathematics books sometimes give addition, multiplication, and exponentiation rules for ∞, such rules do not seem to me to be appropriate. To say that ∞ + ∞ = ∞ does not make much sense, for the opportunity to add infinities does not emerge and if it did there is no definition telling you how to do it.

Pursuing this idea a little further, it is necessary to speak briefly of the limit of a function, such as the limit of the function $1/t$ as $t \to 0$, which is ∞. A function is any rule that gives one specific value for any member of a given set; for example, the function $1/t$ is usually defined for all the real numbers (or perhaps for all the complex numbers) except $t = 0$, and gives values such as 1 when $t = 1$, 1/2 when $t = 2$, and 4 when $t = 1/4$. As noted by Thomas and Finney in their textbook, if $t \to 0$, then $1/t = \infty$, even though $1/t$ is not defined for $t = 0$. The following rules for limits are proved: if L_1 and L_2 are the limits of two different functions, then the limit of the sum of the functions is $L_1 + L_2$ and the limit of the product of the functions is $L_1 \times L_2$, and if $L_2 \neq 0$, then the quotient of the functions is $L_1 \div L_2$. This set of rules is true even when both L_1 and L_2 are ∞, which can be used to make sense of the statements that ∞ + ∞ = ∞, ∞ × ∞ = ∞, and ∞ ÷ ∞ = ∞.

ASSOCIATIONS   When physicists began to study in detail the laws of quantum theory, they were astonished to find that the rules they had adopted soon led to equations whose solutions increased without bound for situations that exist in the real world. For a mathematical function to increase without bound is not a real problem, since mathematics is not necessarily physically real for incommensurables or for various forms of infinity. Not only were these infinities, which physicists call singularities, embarrassing, but also they did not match the physical measurements being determined in laboratories. After a decade or so of confusion, nearly a half-dozen physicists working together and separately developed a way around the problem, a method for quantum electrodynamics called *renormalization*. The standard way to explain this method is to say that some calculations lead to $+\infty$ while others lead to $-\infty$, but when you add together all the results, the various $+\infty$ answers and $-\infty$ answers do not quite cancel out, and the residue is exactly the correct answer that is found in the laboratory when the experiment is performed, correct down to as many decimal places as anyone has been able to check. Richard Feynman, who was among the group that developed the renormalization approach, has suggested that perhaps the incorrect answers of ∞ occur because in the real world, two entities can never be as close to each other as 0, a suggestion that makes sense to me.

When Karl Schwarzschild in 1916 investigated Einstein's theory of gravitation, which is known as the general theory of relativity, other physicists who learned of his results were annoyed in this case also, because his equations persisted in giving ∞ for situations that might occur in the real world. At first, no one was willing to accept these singularities, which were predicted during the collapse of a massive star (even Einstein argued that the solution of ∞ could not make sense in the real world), but starting in 1958 physicists began to make some physical sense of what a solution of ∞ might mean for gravitational theory, a problem on which they worked through the 1960s. By 1967 the concept of a physical entity that corresponds to a mathematical singularity had become sufficiently clear that John Wheeler [American, b. 1911] could name the entity a "black hole." Even though the mathematics was clear, many astronomers believed that it would not be possible for a star to collapse into a black hole; but in 1972 the first likely candidate for a black hole was located. Today it is thought that black holes are in the centers of most galaxies and in some cases may form as supernova remnants. Stephen Hawking [English, b. 1942] has contemplated small black holes that may pervade the universe.

At the heart of every black hole, there is a singularity (∞, in other words) that is matter condensed to such a small radius (not most likely 0, however) that gravity becomes too strong even for electromagnetic radiation, such as light, to escape—that's what makes the hole "black." In practice, however, at a definite distance from the black hole, the gravitational field, although still extraordinarily strong, will allow electromagnetic radiation to escape, although matter in that vicinity will still be sucked in. This radius is called the *event horizon* for the black hole. Gases near the event horizon may be pressured into emitting a lot of electromagnetic radiation that, as normally happens, expands at the "speed of light." This radiation, in some theories, is what we observe as a quasar, provided the black hole is a big one. However, Hawking's little black holes, if they exist, are invisible.

When various reasonable equations for the beginning of the universe are produced, they tend to show a singularity— again, ∞—at the origin of space and time. In other words, ∞ is what precedes the famous big bang that current cosmological theory claims as the origin of the universe. The actual meaning of this ∞ is different in various theories, although most theories simply fail to explain it in any way. I once spent the better part of a year trying in vain to show that the ∞ before the big bang corresponded to a situation in which mass was represented by imaginary or complex numbers instead of real ones, but this idea did not produce any sensible calculations. A theory that I like these days states that before the singularity at the big bang itself, a vacuum existed in which objects of negative mass (or, more accurately, of negative gravitational attraction) spontaneously appeared. These negatives caused the mass to increase (negatively) until it reached the singularity, when positive mass and the big bang arose. Sounds pretty weird, but the mathematics is less weird than the physics. Notice that in this case the singularity is $-\infty$ and not $+\infty$.

Limits

The original idea of what is meant by a limit might be expected to be that something happened "at infinity." But from early times, careful scientists avoided that idea. When Archimedes computed the value of π, for example, he did so with an increasing area of inscribed polygons and a decreasing area of circumscribed polygons. The area of the circle must be between these two extremes—but he did not let the number of sides of either polygon "go to infinity" to find the actual area of the circle. Nearly 1800 years later, in 1585, the problem was addressed by the engineer Simon Stevin [Dutch, born in Flanders; 1546–1620], also known as Stevinius, who wrote on number theory as well as on practical problems of fortification and navigation. Stevin used what amounted to a limit concept in calculating the center of gravity of a triangle but, somewhat like Archimedes', his reasoning was that the center was between two amounts whose difference could be made as small as one wants it to be. Fifty years later Bonaventura Cavalieri [Italian, c. 1598–1647] and Pierre Fermat were calculating areas by adding infinite numbers of infinitesimals. Fermat, however, justified his work by saying that it could be cast in the same method as Archimedes had used for the circle, but hinted that it was too boring to go into that kind of detail. About another fifty years later, in 1687, Newton's *Principia* was published. Although Newton derived his results by a system based on a loose interpretation of limits, he felt that it was necessary to provide all his proofs in the form of the kind of geometry that Archimedes had employed, known as the method of exhaustion. When Newton did publish a work using limits more directly, he warned other mathematicians to take great care in their use.

Well, that did not happen. The eighteenth century is filled with people finding limits right and left by adding infinitesimals and rushing off to infinity. Although there was a lot of wonderful mathematics, sometimes very strange results appeared that were patently false.

Finally, in the middle of the nineteenth century, caution prevailed over adventure. Karl Weierstrass [German, 1815–1897] was only one of several mathematicians whose work made the difference, but his views and notation are the ones that swept the day. Weierstrass defined a limit in terms of finite numbers in a way that is still used today in beginning calculus textbooks (and advanced ones, too, for that matter). Because it is expressed in terms of numbers labeled with the Greek letters epsilon (ϵ) and delta (δ), it tends to be called the epsilon-delta definition of a limit. Sometimes critics who think (perhaps correctly) that students fail to obtain any insight from this careful approach refer to it scornfully as "epislontics." Roughly speaking, the limit is defined by a statement such as "Given any number x and any positive number ϵ, there is a positive number δ such that the absolute value of the difference between a given function and a number L is less than ϵ whenever the absolute value of the difference is less than δ. In that case L is the limit of the function at x." No $\rightarrow$ ∞ or going to 0 is mentioned.

The epsilon-delta idea of a limit can be made somewhat more understandable with the concept of a neighborhood, which in mathematics is a small circle around a point. Generally speaking, one says that a limit of a function is a point such that any neighborhood of the point, no matter how small, encloses at least one point of the function. This definition rests on a theorem of Weierstrass, proved in secret some years earlier by Bernhard Bolzano [Czech, 1781–1848], and now known as the Bolzano-Weierstrass theorem: a set that has bounds and that contains an infinite number of elements must have at least one limit.

Hah! Just when we thought we had banished it, infinity sneaks back in.

$\aleph_0$ Aleph-Null   Denumerable Infinity

FIELD MARKS   The infinity $\aleph_0$ is the completed infinity of all the natural numbers. The choice of the Hebrew letter aleph instead of the more common Greek or Latin letter was that of the mathematician Georg Cantor [German, born in Russia; 1845–1918], whose ancestors were Jewish, although both his parents were Christian. Cantor was the first to investigate infinity in a systematic and thorough way. The other name for this infinity is denumerable—that is, countable—infinity.

Areas of science in which many new terms appear, especially if they are for entities already described by common English words, always seem shaky to me; when the concept is understood, the terminology seems to become simpler. Before plate tectonics was understood, for example, geology was filled with terms like "geosyncline" and "isostasy" that meant little to anyone not initiated to the jargon; after the conceptual breakthrough, the terms used were "plate," "trench," "seafloor spreading," and so forth, words with ordinary meanings. Thus, I suspect that the word "denumerable" stems from a period of uncertainty in dealing with Kingdom Infinity.

That the natural numbers have no highest number was surely recognized very early in the development of mathematics, although Archimedes seems to have been the first to develop a system for representing numbers as great as one might wish to write (*see* discussion at **Genus Natural,** 10^{63}). Archimedes used his system to create a symbol for what we might write as 10^{63}, the sum that represented the number of grains of sand that could completely fill the largest possible universe available to the most speculative of astronomers of classical times. This is much greater than the number of protons, often taken to be about 10^{30}, thought by present-day physicists to be in our universe. Since every physical entity made from atoms must contain at least one proton, it is clear that the total number of the natural numbers that one can envision far exceeds the number of physical entities in the universe. Archimedes' large number, however, is less than the 10^{76} taken to be the total number of particles of all kinds in the universe.

Thoughts of this nature caused philosophers and mathematicians generally to reject the idea that there could ever be a complete list of natural numbers, which is often expressed by saying that no completed infinity can exist. Aristotle specifically excluded the possibility of a completed infinity. Although Euclid is often credited with proving the theorem that there is an infinity of prime numbers, the actual proof states that there is no highest prime number, a different conclusion. (More specifically, Euclid showed that saying that calling some number the largest prime would lead to a contradiction—*see* **Prime Family,** pp. 35–38) Euclid, like other ancient mathematicians, avoided any statement that a completed infinity might exist (*see* discussion at ∞). Philosophers who concerned themselves with what the concept of number might mean found that they could not accept a completed infinity either. Even if numbers are ideas and not objects, how could a finite brain produce an infinite number of ideas?

Galileo [Italian, 1564–1642] demonstrated that if a completed infinity could exist, then the completed number of natural numbers must be the same as the completed number of various sets of natural numbers that do not contain all the natural numbers. For example, the even numbers or the odd numbers each include only half the totality of natural numbers, but for every natural number there is exactly one and only one even number or odd number. For example, the odd numbers can all be expressed by the formula $2n - 1$ where n is a natural number. For each natural number n there is exactly one odd number, which is often shown as a matching like the following:

$$
\begin{array}{ccccccccccc}
1 & 2 & 3 & 4 & 5 & 6 & 7 & 8 & 9 & 10 & \ldots n \\
\downarrow & \downarrow & \downarrow & \downarrow & \downarrow & \downarrow & \downarrow & \downarrow & \downarrow & \downarrow & \downarrow \\
1 & 3 & 5 & 7 & 9 & 11 & 13 & 15 & 17 & 19 & \ldots 2n - 1
\end{array}
$$

Since the odd numbers are evenly spaced, skipping a unit between each two adjacent numbers on the line instead of skipping none, it might be useful to consider increasing the distance between the numbers. If you simply are counting equal-sized spacings, it is like counting a set by twos instead of by ones; clearly, counting by twos produces the same infinity as counting by ones does.

If you increase the distance between members of the lower set of numbers by the next odd number each time, the resulting sequence is the sequence of square numbers. But this sequence can also matched with the natural numbers, since for each number n there exists n^2:

$$
\begin{array}{ccccccccccc}
1 & 2 & 3 & 4 & 5 & 6 & 7 & 8 & 9 & 10 & \ldots n \\
\downarrow & \downarrow & \downarrow & \downarrow & \downarrow & \downarrow & \downarrow & \downarrow & \downarrow & \downarrow & \quad\downarrow \\
1 & 4 & 9 & 16 & 25 & 36 & 49 & 64 & 81 & 100 & \ldots n^2
\end{array}
$$

Instead of being 1 unit apart as the natural numbers are, or 2 apart as the odd numbers are, the distance between each successive pair of square numbers increases by the next odd number (*see* discussion at **Genus Natural, 4**). Looked at this way, the sequence of square numbers is derived from the preceding sequence of odd numbers. Furthermore, this process can be extended. Suppose that the distance between succeeding numbers is increased by the square each time. No matter how sparse the sequences become by this method, they continue to match with the natural numbers.

It was Dedekind who suggested using this property as a definition of a completed infinity. Following his lead, the most common contemporary definition of an infinite set is one that can be matched with a proper subset of itself. Technically a proper subset is any subset that does not contain all the members of the original set, but for infinities the proper subset used in discussion often fails to include very many members of the main set.

SIMILAR SPECIES When a sequence is taken to infinity, one writes $n \to \infty$, using the symbol for "increases without bound," for which n by convention is considered to be a natural number. Thus, in this case, the totality of terms of the sequence is $\aleph_0$, the denumerable infinity, since each member of the sequence is associated with a specific natural-number value of n. But there are other reasons why ∞ is clearly different from $\aleph_0$. Often the increase without bound is not confined to the natural numbers, and thus the totality of terms is different from $\aleph_0$.

Even in the light of the discovery of other infinities, it is still surprising to learn that the infinity of rational numbers is equal to the infinity of natural numbers, that is, like the natural numbers, the rational numbers are countable. It is surprising that there are $\aleph_0$ rational numbers because no two rational numbers are next to each other—there is always another rational number, indeed an infinity of other rational numbers, between any two different rationals. For example, between 2 and 3 there are 2.1, 2.2, 2.3, and so forth; between 2.1 and 2.2 there are 2.11, 2.12, 2.13, and so forth; and between 2.11 and 2.12 there are 2.111, 2.112—but you get the idea.

When Cantor first showed that the infinity of rationals was the same as the infinity of naturals, the idea that there might be more than one infinity was unknown to most people and not clear even to Cantor himself. Soon, however, Cantor showed that the infinity of real numbers was different from that of the naturals and rationals, and he also developed a method for producing one infinity from another.

Before tackling the rational numbers, however, let us look at a simpler problem, which is counting the number of integers. It is easy to see that this can be done simply by taking integers farther and farther from 0 on each side.

350

$$\begin{array}{ccccccccccccc} 1 & 2 & 3 & 4 & 5 & 6 & 7 & 8 & 9 & 10 & \dots & 2n & 2n+1 \\ \downarrow & \downarrow & \downarrow & \downarrow & \downarrow & \downarrow & \downarrow & \downarrow & \downarrow & \downarrow & & \downarrow & \downarrow \\ 0 & 1 & -1 & 2 & -2 & 3 & -3 & 4 & -4 & 5 & \dots n & & -n \end{array}$$

Thus, the integers are also countable.

The American philosopher (and founder of pragmatism) and mathematician Charles S. Peirce [1839–1914], son of Benjamin Peirce, in 1883 suggested a way of arranging the positive rational numbers to show that they are countable (equal in size to $\aleph_0$). This method has some advantages over the method used by Cantor.

Peirce begins with the fractions 0/1 and 1/0. Remember that fractions are numerals, so 1/0 simply means 1/0 and is OK. Although the fraction 0/1 represents the rational number 0, the fraction 1/0 does not stand for a rational number; it is also not ∞, as the discussion of potential infinity shows. Eventually 1/0 will be tossed out, since what we are counting are the rational numbers. Thus, the first set contains 1 rational number 0/1 and one device used in construction—a scaffolding to be torn down at the end.

New rational numbers are generated by Peirce's method in much the same way as the main theorem concerning Farey sequences (*see* **Farey Sequences,** pp. 239–242), which is that adding numerators and denominators of two terms out of a triplet of terms will produce the one in the middle. In Peirce's procedure, the new rational number is similarly represented by a numeral in which the sum of the numerators of adjacent fractions becomes the numerator of the new fraction, and the sum of the denominators of adjacent fractions becomes the denominator of the new fraction. If 0/1, 1/0 is called *Step 0* then *Step 1* produces a new fraction $(0+1)/(1+0)$ or 1/1, and the step itself is 0/1, 1/1, 1/0. Proceeding in the same way, *Step 2* becomes 0/1, $(0+1)/(1+1)$, $(1+1)/(1+0)$, 1/0, which then is written as 0/1, 1/2, 1/1,

2/1, 1/0. Continuing with *Step 3,* we obtain 0/1, 1/3, 1/2, 2/3, 1/1, 3/2, 2/1, 3/1, 1/0. *Step 4* produces 0/1, 1/4, 1/3, 2/5, 1/2, 3/5, 2/3, 3/4, 1/1, 4/3, 3/2, 5/3, 2/1, 5/2, 3/1, 4/1, 1/0. Let us call these *Peirce sequences* and say that *Step 4* is the fourth Peirce sequence.

Notice that this procedure will inevitably produce all possible natural-number numerators as well as all possible denominators (since 1 must be added repeatedly, as in the Peano postulates for the natural numbers, where the natural numbers are generated because each number has a successor formed by adding 1). The rational numbers less than 1 in a Peirce sequence are similar to Farey sequences because of the way they are formed, but fail to include all members of a Farey sequence in a given step. The rational numbers greater than 1 in a Peirce sequence are the reciprocals of the numbers less than 1, with the expression 1/0 counting for the reciprocal of 0.

As we continue to higher and higher steps, the numbers between 0/1 and 1/1 crowd together more, while the numbers larger than 1 spread farther apart with each step. Despite this spreading out, it is apparent that eventually any rational number you can name will be a part of a Peirce sequence for *some* step, since all possible numbers from 0 to 1 appear in some step if you continue far enough.

Now comes the hard part. Peirce argued that you can use his sequences to show that the natural numbers can be matched in a one-to-one way with the rational numbers and you can do it without eliminating any fractions because they are not in lowest terms. The construction guarantees that all will be in lowest terms to begin with (except that all rational numbers that are also natural numbers will be written in the form $n/1$ instead of just as n). There may be a way to do this without eliminating duplications that I don't know, but if you allow elimination of duplicate fractions, you can do the matching as follows. Arrange all the Peirce sequences in order from *Step 1* to *Step ∞*, ignoring *Step 0*. By going to *Step ∞*, I mean that you should do as you do with the natural numbers and be aware that these sequences carry on to potential infinity. Now begin matching the natural numbers 0, 1, 2, 3, . . . with the rational numbers as shown by the fractions, skipping over any rational numbers already counted. The result of such a matching would begin as follows:

0 1 2 3 4 5 6 7 8 9 10 11 12 13 14 15...
↓ ↓ ↓ ↓ ↓ ↓ ↓ ↓ ↓ ↓ ↓ ↓ ↓ ↓ ↓ ↓
0/1 1/1 1/2 2/1 1/3 2/3 3/2 3/1 1/4 2/5 3/5 3/4 4/3 5/3 5/2 4/1. . .

As you can see, this process could continue as far as one likes (to ∞, so to speak) and there would always be a one-to-one matching. Therefore the infinity of positive rational numbers is $\aleph_0$, the same as the infinity of natural numbers. It was convenient in this case to start the sequence of natural numbers with 0, but the argument would be exactly the same if we started with 1. To obtain all the rational numbers from the positives, you can use the same procedure that was used to get from the natural numbers to the integers.

PERSONALITY For $\aleph_0$, "infinity plus infinity is infinity." We begin by noting that $\aleph_0 + 1 = \aleph_0$, $\aleph_0 + 2 = \aleph_0$, and so forth, which is easy to see. Then if follows that $\aleph_0 + \aleph_0 = \aleph_0$. Similarly, $2 \times \aleph_0 = \aleph_0$, $3 \times \aleph_0 = \aleph_0$, leading to $\aleph_0 \times \aleph_0 = \aleph_0$.

An important rule that we will need later is that it is possible to omit an infinite number of elements from a set without changing the infinity of the set.

$\aleph_0$

ASSOCIATIONS The concept that the rational numbers are in one-to-one correspondence with the natural numbers—that is, that they are countable—suggests that many other sets thought to be somehow of different sizes are also countable. It should be clear that only mathematical objects can be both infinite and countable, but there are many possible mathematical objects that can be studied. Among them, the algebraic numbers (*see* **Algebraic Family,** pp. 268–271) are also countable. The explanation of this begins with a somewhat different way of showing that the set of rational numbers is countable. Instead of rational numbers as fractions, consider the following array of ordered pairs, which extends to ∞ both to the right of the list and also to ∞ as one continues down the page:

$$
\begin{array}{lllll}
(0, 0), & (0, 1), & (0, 2), & (0, 3), & (0, 4), \ldots \\
(1, 0), & (1, 1), & (1, 2), & (1, 3), & (1, 4), \ldots \\
(2, 0), & (2, 1), & (2, 2), & (2, 3), & (2, 4), \ldots \\
(3, 0), & (3, 1), & (3, 2), & (3, 3), & (3, 4), \ldots \\
(4, 0), & (4, 1), & (4, 2), & (4, 3), & (4, 4), \ldots \\
\ldots
\end{array}
$$

If you follow the arrows (and picture the process continuing), you can see how to count the ordered pairs. If what you want to do is to prove that the rational numbers are countable, then you eliminate any ordered pairs that are equal to ones you have already counted. But for algebraic equations of the form $ax + b = 0$, where a and b are natural numbers (including 0), this list would also include all the possible values of a and b. Now the same kind of counting can be extended to any ordered ntuple of numbers. First, extend to the ordered triples by making the first member of each ordered pair an ordered pair of natural numbers itself, while the second is a natural number. Since the ordered pairs of natural numbers are countable, this can clearly be accomplished, so the ordered triples are also countable. Then use the triples as the first members to show that the ordered quadruples are countable. And so on. While you must stop before the totality of elements in an ntuple becomes infinite, this approach will work for any finite n. Similarly, the ordered ntuples become the coefficients of any sized algebraic equation, just as a and b were the coefficients for the simple linear equation. Although each algebraic equation can have as many solutions as the degree of the equation, these solutions can also be arranged in ntuples. Thus the algebraic numbers are countable. Notice that this proof applies to the complex algebraic numbers and not just the real ones.

I can't resist mentioning a second way that has been proposed to show that the algebraic numbers can be shown to be countable. Every algebraic equation is a combination of just fourteen symbols (provided that exponents are written on the line, so that an equation such as $2x^2 + 3x - 5 = 0$ is written as $2x2 + 3x - 5 = 0$). The fourteen symbols are the 10 digits of the decimal numeration system along with x, $+$, $-$, and $=$. Call the symbols digits in a numeration system with base 14, so that x is 10_{ten}, $+$ is 11_{ten}, $-$ is 12_{ten}, and $=$ is 13_{ten}. Then in that system the number 25_{ten} would be written as $1+_{fourteen}$ and 97_{ten} would be $6=_{fourteen}$, and so forth. Although not every number in base 14 would be a sensible algebraic equation, every sensible algebraic equation would be the numeral for a natural number in base 14. Thus, the algebraic equations (and therefore the algebraic numbers) are countable.

c $\aleph_1$ $2^{\aleph_0}$ The Power of the Continuum

FIELD MARKS   Richard Dedekind put it this way: "The straight line L is infinitely richer in point-individuals than the domain R of rational numbers [is rich] in number-individuals." In other words, there are far more points on a line than can be matched with the rational numbers. It is not clear how Dedekind reached this conclusion, since this was two years before Georg Cantor's first proof of this result and many years before Cantor's most famous proof that there are more real numbers than rational numbers, the one that is generally cited in discussions of this topic.

Georg Cantor, who investigated infinite sets quite thoroughly around the turn of the nineteenth into the twentieth century, showed that for any set with an infinity of members there must be a greater set that consists of all the possible subsets of the infinite set. This is a generalization of the property for finite sets that a set with n members has 2^n subsets. A subset of a set S is simply any set for which all its members are also members of S. A careful examination of this definition shows that among the subsets of S are the set S itself and also a set that has no members, called the null set or empty set. A finite example of a set and its subsets would be the set {A, B, C}, which has three members and also has the subsets {} (the empty set), {A} (the set with A as a member), similarly {B} and {C}, {A, B}, {A, C}, {B, C}, and {A, B, C}. Notice that there are $8 = 2^3$ subsets, as predicted. The set of all subsets of a set is often called the power set of the original set, or $P(S)$ for a given set S.

Here is a sketch of the proof that the power set always has more subsets than there are members of the original set. Suppose that each element of S is matched with one and only one of the subsets in $P(S)$. If this matching includes all the subsets, then the S and $P(S)$ have the same number of members. In the matching, there are two possibilities to consider. The element from S can either be a member of the subset of $P(S)$ or not a member. Now consider a set composed of all the members of S that in this matching are not members of the subsets with which they are matched. This set is clearly a subset of $P(S)$, but it is not matched with any of the members of S. Therefore there are more sets in $P(S)$ than there are members of S. Notice that infinity was not mentioned, but there is nothing in the proof that prevents S from being an infinite set, such as the countable set of natural numbers. In that case, the infinity associated with $P(S)$ is greater than $\aleph_0$, the infinity of the countable sets.

Notice how close the proof of this theorem is to a famous paradox about sets discovered by the English philosopher, mathematician, and logician Bertrand Russell [1872–1970]. Russell defined two groups of sets: normal sets that are not members of themselves and abnormal ones that are. The paradox arises when one asks whether the set of all normal sets is itself normal or abnormal. The contradictions that result in this case are among the most difficult to eliminate from the theory of sets, and some would say can be eliminated only by such radical methods as not permitting discussion of certain kinds of infinite sets at all. There are other proofs of Cantor's theorem that I have seen, but all the ones I know are very close in structure to Russell's paradox. This casts something of a shadow on the further discussion of infinity.

The next higher infinite number beyond the infinity of the natural numbers, designated as $\aleph_0$, is given the next subscript, so it is $\aleph_1$. It is also the same as 2 to the $\aleph_0$ power. Although $\aleph_0$ and $\aleph_1$ are both infinities, you should know that when you omit $\aleph_0$ elements of a set with $\aleph_1$ elements, the left-over set still has $\aleph_1$ elements. (The proof of this uses mathematical induction, as in the Peano postulates (*see* **Genus Natural**), but would take several pages to explain in a step-by-step fashion. For once let me say "trust me.")

Cantor was able to show that the number of subsets of the infinite set of natural numbers, or $\aleph_1$, is the same as the number of members of the real numbers, which he designated c in honor of the real number line, which is known as the continuum. (This c in mathematics has nothing whatsoever to do with c in physics, the speed of light in a vacuum, most familiar from Einstein's equation $E = mc^2$.) Another name for c is "the power of the continuum." Thus $\aleph_1 = c$ and therefore c is greater than $\aleph_0$. The question naturally arises, is there any infinity between c and $\aleph_0$? A statement of the negative answer to this is known as Cantor's continuum hypothesis. Mathematicians have proved that there is no way of establishing whether or not this continuum hypothesis is true, since both the hypothesis and its opposite are consistent with the rest of mathematics as it is currently formulated. Like Russell's paradox, this is somewhat of an embarrassment, but also like Russell's paradox there seems to be no easy way around it. Logicians who have dealt with this issue rather think that at some time in the future someone will find a new basis for mathematics that will avoid such an apparent flaw, but so far no one seems to have managed the trick.

There are two ways to approach the infinity of c. One is to show that by itself it is uncountable, a proof that is known as the diagonal method. The other is to show that it is the same as $\aleph_1$, a method that is less often described in popular accounts of infinity, but one that is less unsettling. Here is a brief sketch of the second method: show that every subset of the natural numbers, which for all the subsets has infinity $\aleph_1$, can be described by the binary representation of a real number between 0 and 1, and that all such numbers, which have infinity c, are represented.

The diagonal method is much more commonly quoted, although it is an indirect proof, one that proceeds by contradiction. First one must suppose that the set of real numbers is countable. If so, then a proper subset, such as the real numbers from 0 to 1, will also be countable, perhaps by some scheme in which the order is changed as was done for counting the rational numbers (*see* discussion at $\aleph_0$). In that case, all the infinite decimals representing the numbers from 0 to 1 can be arranged in order starting with the first and continuing on. Notice that each decimal for a real number also has countably many infinite digits.

The diagonal method is to take the first digit from the first decimal in the arrangement, the second from the second, and so on, always taking the nth digit from the nth decimal when they are arranged in the countable sequence. Each digit is then changed in some way to a different one—this can be done systematically by adding 1 to each digit that is not 9 and changing 9 to 5, for example. A new decimal is constructed that has the changed first digit followed by the changed second digit and so forth, so that the nth digit in the new decimal is the changed nth digit from the nth decimal in the arrangement. But this new decimal represents a real number between 0 and 1 that was not in the original arrangement that was to have contained *all* the decimals between 0 and 1. It is clear that it is not the same as the first decimal since it differs from it in the first place; it is not the second, because it is different in the second place; and, in general, it is not the nth decimal, since it differs from that one in the nth place. Hence, there is a real number that is not in the arrangement of all the real numbers between 0 and 1 in countable order, and therefore the real numbers cannot be countable.

Some people are uncomfortable with the diagonal method, although it is the basis of many of the important proofs dealing with the foundations of mathematics. For one thing, it depends on being able to choose an infinite number of entities from an infinite number. It can be shown that such an operation depends ultimately on a principle called the axiom of choice, which at its heart simply says that it is possible to choose specific elements from a collection of infinite sets without specifically constructing each element. If the axiom of choice is true, then one has to accept the diagonal method as a reasonable proof. If not, then there are many adjustments that need to be made in one's view of what mathematics really is and what it can do. Thus, most people choose to accept the axiom of choice.

The proof as stated has one other possible loophole, which is that in some circumstances two different infinite decimals represent a single real number. If the rule includes changing 9 to 0, instead of the suggested 5, then this can be a problem, for in base ten a number ending in an infinite sequence of 0s becomes the same as another number ending in an infinite sequence of 9s. But notice that the method suggested will fail to produce any decimal with the digit 0 in it at all and, although an infinite sequence of 8s could become all 9s, that would mean the same real number as any one ending in an infinite sequence of 0s—but the new decimals have no 0s at all.

Getting from the interval from 0 to 1 to the entire set of real numbers may strike you as another difficulty, but this is easily accomplished with geometry. Consider a triangle that has as its base the interval from 0 to 1. Now extend the other two sides of the triangle beyond that base as far as you wish—to ∞, as it were. Draw the real line—that is, a line with all the real numbers on it—so that it meets the two extended sides of the triangle at some distance from the base. Choose any real number on the real line and connect it with the vertex of the triangle that includes the extended sides but not the base. The line connecting the vertex and the chosen number will pass through the base of the triangle at one particular point, which must be matched with one of the real numbers between 0 and 1. Here is a sketch showing how this procedure could be used to match, for example, π with one of the decimals representing a number between 0 and 1.

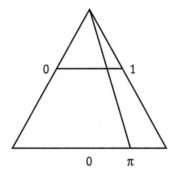

Notice that if you need to reach some real number that is beyond the end of the real line between the extended sides, you need only move the real line farther from the vertex and, if you move it far enough, you can reach any real number that you might desire.

SIMILAR SPECIES In recent years the seekers in the foundations of mathematics have discovered that very large infinities are important in defining the strength of mathematical theories. One concept is that of the "inaccessible cardinal," an infinity that cannot be derived from c or from the sequences of $\aleph$s. The infinities in the sequence of $\aleph_0$, $\aleph_1$, and continuing on are called *cardinal infinities*, since the numbers used to count—1, 2, 3, and so forth—are the cardinal numbers. Some other large cardinals not in the sequence of alephs have been characterized, despite inaccessibility, in

c

various ways. True to tradition in mathematics, some have been given innocent-sounding names that have complex definitions behind them, such as huge cardinals (my favorite), measurable cardinals, ineffable numbers, and weakly compact cardinals. Others are named for the mathematicians who investigated them, as in Mahlo or Ramsey cardinals. These are all cousins of c, of course, but the details are more than any of us care to know.

You might think that once you have shown that the number of points on a line, or c, is greater than the number of counting numbers, or $\aleph_0$, you would then calculate that the number of points on a plane or in 3-space will be greater than c. The oddity of infinities comes into play, however. Cantor was surprised to find a proof that the number of points on a plane is the same as the number on a line (remember that Dedekind's definition of an infinity is a set that can be put into one-to-one correspondence with a part of itself). The proof is much simpler and more direct than the diagonal method. It goes like this.

Any point on a plane can be represented as an ordered pair of real numbers. Any real number can be represented by an infinite decimal. Consider a typical point (x, y) of the plane. It is easier to begin with a point in the unit square that has both coordinates between 0 and 1. To construct a single infinite decimal from the countable infinity of digits in x and y, start with a decimal point and then put the first digit of x first, then the first digit of y, then the second digit of x, then the second of y, and continue on this way. When you have used all the countable infinity of both sets of digits, you will still have a countable infinity, and therefore the single new decimal constructed from the digits in the pair of coordinates will also be a real number between 0 and 1. This shows that you can map all the points in the unit square into the part of the real line between 0 and 1 in such a way that every point on the square corresponds to exactly one point on the line.

You can go the other way just as well. Start with a point on the real line between 0 and 1. To construct a point in the unit square, begin with two decimal points. After the first you put all the odd-numbered digits from the representation of the real number. After the second you put all the even-numbered digits (beginning with the second). The two numbers thus defined are a pair of real numbers that are the coordinates of a point in the unit square. Thus, every real number on the line corresponds to exactly one point on the unit square. This completes the requirements of one-to-one correspondence, so there are exactly as many points on the line as in the unit square. From this relationship there are a number of ways to show that the same applies to all the points on the line and all the points in the plane.

Furthermore, by choosing the first, second, and third digits, and so forth, you can map the points on the line into 3-space and the points on 3-space into the line. Indeed, the same procedure would work for space of any specific natural-number number dimension. I needn't go into this further, but the same ideas can even be extended to spaces with infinite dimensions, such as $\aleph_0$-dimensional space. (If you want to work it out for yourself, you use the same twisting path that was used to show that the algebraic numbers are countable.)

At this point, it might be easy to stop, thinking that all sets are either countable or have the infinity c of the real numbers. But recall that the power set of a set always has more members than the set itself does. If we consider the power set of the real numbers, it must have more members than c. This number would be designated in the aleph structure as

$\aleph_2$. Most of the entities that have the infinity $\aleph_2$ of members are sets of subsets, but one set with this infinity of members is a familiar set in its own right. It is the set of all functions of real numbers. The proof is another easy one. Each function of real numbers can be represented on the plane by a subset of the plane. There are $\aleph_2$ subsets of the plane (since the set of points on the plane has cardinal number c), so at best the number of real functions could be no greater than $\aleph_2$. But a subset of all the real functions are those that "pick out" subsets of the real line by mapping every real number into one of the subsets of the real line. There are at most $\aleph_2$ such functions, since that is the number of subsets of the real line. Thus, the number of functions must be exactly $\aleph_2$.

PERSONALITY   One of the properties of infinite sets is the following: if the cardinal number of a set is greater than $\aleph_0$, then that number will be unchanged by the removal of any countable subset of elements. One of the common applications of this in the theory of real functions is to remove all the points represented by rational numbers from the real line, leaving only the irrationals. The remaining set will still have the cardinal number c. Indeed, even if you remove all the points from the line whose coordinates are algebraic numbers, leaving only the transcendentals, the number of points remaining will still have the cardinal number c. Since we know only a few transcendental real numbers (π and e are about the only two that most people ever encounter in their whole lives), the feeling that this procedure produces is that we have taken nearly all of the line away, so it seems strange that there are more points left than have been removed. But the feeling is wrong, for there are more transcendental numbers than algebraic ones. We just don't know specifically what most of them are.

ASSOCIATIONS   The foundation of Feynman's theory of particle physics is that one assumes that a particle takes all possible paths to get from one place to another, which I take to be *c* possible paths. This physical theory has a sort of analog in the theory of complex functions. Gauss was the first to note (in 1811, although he did not offer a proof) that in integration of a complex function all possible paths between two points—complex numbers—give the same correct result *unless* the function has an infinite value somewhere along the way. A few years later Cauchy produced a proof and also found a way of solving the case where the function becomes infinite.

ω   First Ordinal   First Transfinite Number

FIELD MARKS   When Georg Cantor established that some infinities are greater than others, the notion greatly disturbed many of the mathematical minds of his day, who said it was nonsense and there could be no grounds for it. He solved their philosophical dilemma by pointing out that numbers serve two functions, ordinality and cardinality—that is, size and sequence. Generally these two go hand in hand: 10 is greater than 1 and it comes after 1. But they need not be linked, and according to one theory, it was this idea that was hampering his colleagues in understanding how some infinite sets could be larger than others. For me, however, there is no problem dealing with the aleph series of infinite numbers. But when I get to the infinity represented by the ordinal numbers, I always find that I go weak in the mind. Nevertheless, for the sake of completeness, here is a brief summary of the ordinal situation as it passes beyond the finite (and beyond most of my finite grasp of it).

The ordinal number ω is the first extension of the set of natural numbers, not the cardinal infinity of natural numbers (which is $\aleph_0$). Thus, when counting you begin 1, 2, 3, ... and eventually come to ω, $\omega + 1$, $\omega + 2$, $\omega + 3$, ... The way that this works is that the idea of one-to-one correspondence, which is the basis of the cardinal infinities, is replaced with the idea of matching two sets so that the matching preserves the order of both sets.

The definition of ω often given is that it is the ordinal number of the ordered set of natural numbers, but ω is better defined as the smallest number greater than any finite ordered set of natural numbers. As such, it is properly called a transfinite number, not an infinity. Although I have read numerous discussions of ω, the actual line of reasoning that establishes that ω must exist eludes me. The method of going from the natural numbers in order to get the next number after them is called transfinite induction, but it just seems that you say "Well, here are all the natural numbers in order. When you finish counting them, then you go on to the next numbers, which are ω, $\omega + 1$, and so forth. That's transfinite induction."

Sometimes Cantor's continuum hypothesis is stated as the assumption that ω is the same as c, which cannot be proved. This seems to equate a cardinal infinity with an ordinal transfinite, however, which should not work.

SIMILAR SPECIES Cantor defines a number ϵ_0 that is equal to the limit of ω taken to the ω power taken to the ω power and so forth for a total of ω times. He goes on from there to produce versions of ϵ with progressively greater subscripts. In any case ϵ_0 is the smallest infinity that you cannot obtain by adding, multiplying, or even taking powers of a finite number of infinities.

PERSONALITY   As a result of this odd definition of numbers beyond ω, $1 + \omega = \omega$ but $\omega + 1 > \omega$. Thus, addition with ordinal infinities is not commutative. Remember that we are talking about sequence, not size, here, so it should not be so surprising that addition is noncommutative. First before second is different from second before first. This applies to adding the ordinal infinities themselves as well as adding the counting numbers to ω. If you add $(\omega + 2) + (\omega + 1)$ the result is written as $\omega + \omega + 1$, but if you add $(\omega + 1) + (\omega + 2)$ the sum is $\omega + \omega + 2$.

For multiplication the rules are somewhat like those for $\aleph$ but not the same. For example, $2 \times \omega = \omega$, not 2ω. Multiplication, however, is not commutative either, so $\omega \times 2 = \omega + \omega$.

The rest of the arithmetic of ω is a mix of the familiar and the strange. The natural-number powers of ω are just like those of a real number; for example, $\omega^3 = \omega \times \omega \times \omega$. But there is also a power ω^ω, which equals $1 + \omega + \omega^2 + \omega^3 + \ldots$

ASSOCIATIONS   Nonstandard analysis is a branch of mathematics that since 1966 has tried to develop a sensible interpretation of what it means to be an infinitesimal, or infinitely small number. Although not everyone would agree, I observe that the infinity used in nonstandard analysis is essentially the same as ω. To confuse the issue a lot, some mathematicians use ω to mean the reciprocal of infinity. That is, they define a number ν to be greater than any of the natural numbers, and set ω equal to $1/\nu$. Given all the possible symbols one can use, this choice seems almost perverse!

Despite these kinds of problems in terminology, people who have studied nonstandard analysis have concluded that proofs of the familiar results from standard calculus are just as rigorous but easier the nonstandard way, so the day may come when the nonstandard approach is taught to college freshmen. Furthermore, there are even a few results from nonstandard analysis that are important and that had not been discovered using the ordinary methods. Thus, the actual infinite and even the actual infinitesimal may be the mathematics of the future, replacing the fear of the actual infinite that has persisted from Aristotle through Gauss and down to the present for many of us—depending, in some cases, on which infinity you are talking about.

Further Reading

The following books contain additional information on numbers, mental computation, and infinity.

Numbers

Clawson, Calvin C. *Mathematical Mysteries: The Beauty and Magic of Numbers*. New York, Plenum Press, 1996.

Conway, John H., and Richard K. Guy. *The Book of Numbers*. New York, Copernicus (Springer-Verlag), 1996.

Dehaene, Stanislas. *The Number Sense*. New York, Oxford University Press, 1997.

Guedj, Denis (tr. by Lory Frankel). *Numbers: The Universal Language*. New York, Harry N. Abrams, Inc. 1997.

Menninger, Karl (tr. by Paul Broneer). *Number Words and Number Symbols: A Cultural History of Numbers*. New York, Dover, 1992.

Ore, Oystein. *Number Theory and Its History*. New York, Dover Publications, 1988.

Phillips, Richard. *Numbers: Facts, Figures, and Fiction*. Cambridge, MA, Cambridge University Press, 1995.

Wells, David G. *The Penguin Dictionary of Curious and Interesting Numbers*. New York, Penguin USA, 1998.

Mental Computation

Julius, Edward H. *Arithmetricks: 50 Easy Ways to Add, Subtract, Multiply, and Divide Without a Calculator*. New York, John Wiley & Sons, 1995.

Infinity

Lavine, Shaughan. *Understanding the Infinite*. Cambridge, MA, Harvard University Press, 1998.

Morris, Richard. *Achilles in the Quantum Universe: The Definitive History of Infinity*. New York, Henry Holt & Company, 1997.

Pickover, Clifford A. *Keys to Infinity*. New York, John Wiley & Sons, 1997.

Number Index

Numbers with interesting properties discussed in this Field Guide are indexed below in the order of their absolute values; when absolute values are the same, positive numbers are listed first, followed by negatives and complex numbers.

Word Index

tetrahedron, 60
Thales, 52
Theaetetus, 263
Theory of Everything, 54
Thirty-six Officers Problem,
    150–151
Thomas, George B., Jr., 334–335
thousand, 101
three-space, infinity, 363
Three Theological Virtues, 85
time, 197–198
tip, 140
tithe, 220
Titius, Johann Daniel, 148
tokens, 12–15
torus, 73
Transcendental Family, 290–292
transcendental number, 1, 143,
    285, 290–292, 364
transfinite induction, 366
transfinite number, 365–368
transformations, 339–340
transiterate, 177–179
trefoil knot, 43
triangle, 30, 41, 75
Triangular Family, 44–47
triangular number, 1, 28, 39,
    44–47, 81, 102, 129, 148,
    241
  formula, 47
triangulation, 123–124
trigonometric function, complex
    number, 324
Trinity, 41
trisection of angle, 41–42, 291
trivial, 42

trivium, 42
truss, 151
Tsu Ch'ung-Chih, 294
Tsu Keng-Chih, 294

unary operation, 193
unit fraction, 212–213, 216–217
unit vector, 310
universe, 180, 209, 261, 331,
    343–344

van Ceulen, Ludolph, 295
vector
  i, 309–310
  three dimensions, 327
Viète, François, 295

W particle, 209
Wallis, John, 190–191, 295–296,
    334
wasp, 67
wave, regular, 325
weak force, 54
Weierstrass, Karl, 345
Weiske, Adolph, 54
Wheeler, John, 343
Wilson prime, 63, 119
witch's coven, 118
writing numbers, 71
Wu, Chien-Shiung, 197

Xenophanes, 52

year, 158–159, 169

Z particle, 209

Word Index